THE RENEGADE SPY PROJECT

Terri Selting David

D1397615

To my very own Wren, whether she decides to be an engineer, artist, writer, scientist, programmer, or everything all at once.

And to her brother whose limitless patience, creativity, and kindness astound me.

1

TROUBLEMAKER

*W*ren Sterling took a deep breath and hurled herself into the Friday morning chaos.

She'd stood rooted outside the front door of Ada Lovelace Charter School as the other students flowed around her like one lone tree in a flooded river, and now she was late. She needed to make it to Principal Sophie's office for another dumb lecture before the first class. Someone bumped into her from behind, pushing past as the halls surged and boiled with kids.

It was that stupid student election, mostly.

Wren looked around cautiously, trying to find a clear path through the maze of council candidates and other students. The candidates employed their entire arsenal of posters, badges, magnets, and, the most sinister of all, cupcakes. Those cupcakes were *dangerous*. They lured you in with their delicious vanilla frosting, their glorious rainbow sprinkles and — no. Wren could not get distracted by the sugary little fiends. She had to make it through.

A tall kid walked by. Seizing the opportunity, Wren used him as a shield, walking in his wake until he made eye contact with a candidate and stuttered to a stop. Rookie mistake. There was no escaping a campaign speech once you made eye contact. She slipped past, unseen.

Locker magnets, fliers, and baked goods sprang out as she moved down the halls. If she lost focus, even for a second, some form of propaganda would materialize right in front of her. She had to grab whatever it was before colliding into it, face first. Which was especially bad with the cupcakes...

Surfacing at the principal's office without being pulled aside once, she waved at Ms. Sophie's secretary. He smiled at her and indicated it was safe to enter. She took another deep breath and yanked the door open.

Inside the cozy office, Principal Sophie looked up from scribbling on some official looking forms. Wren tossed her backpack on the floor and draped, sideways and dramatic, across the armchair in front of the desk. Her head dangled over one of the chair's arms, her feet over the other.

"I don't want to talk about it," she groaned.

Principal Sophie, a small, middle-aged woman in a fancy suit, sighed. "I hoped to see less of you this year, Wren."

Wren flashed the principal a mischievous smile. "Awww, I thought you liked me!" She was beginning to understand why people didn't sit sideways more often. It was really uncomfortable.

"Tell me about Mr. Vincent's class yesterday," Ms. Sophie prompted.

"Same thing as always," Wren pulled herself into a regular sitting position, then dropped her head onto the desk with another loud groan. Much more comfortable, still properly dramatic. "I try as hard as I can, but my body does things without telling me first. I don't want to cause trouble, but I can't sit still and listen. Birds and cars are loud and the lights are too bright. You know what? Bobby had egg salad for lunch. I could totally smell it because he got some on his shirt. ON HIS SHIRT, Ms. S. and he didn't even care! So disgusting. And honestly, can I tell you a secret? I think Mr. Vincent might actually be a robot. An emotionless robot who doesn't understand the vast complexity and range of human feelings. Either that or he's just a jerk. I don't even know what I'm doing wrong until he tells me! Okay, sure, maybe I was a little loud about expressing my opinion but honestly, what's so wrong about that anyway? Seriously," she gazed up hopelessly into Ms. Sophie's face, "aren't we supposed to stand up for what we believe in?"

"There's a right way and a wrong way to stand up for what you believe in, Wren. You know that," the principal leveled her famous stern expression at Wren. The one that said she wasn't putting up with your shenanigans. The one that stopped most students in their tracks.

But Wren Sterling wasn't most students. It took more than a stern look to intimidate her, no matter how legendary. The entire world was filled with stern looks for a girl like Wren. She'd been on the receiving end of Ms. Sophie's so often since kindergarten that it just didn't have the same power anymore. And she knew that, secretly, the principal of Ada Lovelace Charter School was actually a big softie.

Lovelace, in the middle of San Francisco, was huge. Kindergarten through eighth grade. The elementary school populated the first three floors, while middle school dominated the top two. At just-turned-eleven, Wren was officially the youngest kid in sixth grade. She'd waited forever to be up on the fourth floor, to leave elementary school behind. Make a big change.

This year was supposed to be different. The year she'd get in less trouble. The year everything would change.

But life wasn't magically better up that flight of stairs.

Classes had started three weeks ago, and so far all she'd gotten out of middle school was more exercise.

"You need to find a better outlet for your... enthusiasm. You're a smart girl, Wren. Make better choices." And now Ms. Sophie was lecturing her again, as she'd done for years. Nothing had changed.

Wren looked out the window, slowly shaking her head, "I'm not sure I can. I honestly try so hard. I'm exhausted from trying so hard. But I'm just not like the other kids. My brain works differently, I know it does. The only place it quiets down is when I'm making stuff in my tinkering club. I don't WANT to see you so much," she paused. "I mean, no offense..."

Ms. Sophie's scowl softened at the corners.

"I believe you, Wren, I honestly do," she said. "You've never lied to me, even when the truth got you into more trouble, and I appreciate that. You're a fascinating girl. You just need to, I don't know, tone it down? Be a problem solver, not a problem starter. Okay?"

"Be like everyone else," Wren agreed. "Gotcha."

"That's not quite what I mean..."

"Actually, yeah. That just might work! Be boring like the other kids," Wren nodded to herself, lost in thought. "Brilliant!"

Ms. Sophie's scowl melted completely into a smile. "Thank you for coming in before class. You can go on now. Don't be late. Rules are rules."

As Wren stood and gathered her things, Ms. Sophie added, "And I do like you, Wren. But I don't want to see you so much this year, either."

"I understand. I promise I'll try even harder to be good," Wren called back over her shoulder. "And boring. Totally boring."

The hallways were the most dangerous right before the start of class. Candidates, desperate to get in one last vote-winning speech, became even more aggressive. Kids lost their focus rushing to class. And here it was, T-minus-five minutes. But she wasn't going to get in trouble today. Nope. Not today. She'd promised.

As Wren climbed the stairs to the fourth floor, a boy pressed a small piece of paper into her hand. She grabbed it as she pushed past, his voice disappearing behind her.

"McGuckin for Council! Vote Peter!" the paper said, with a drawing of a dog... or maybe it was an otter? Wren dropped it into the closest recycling bin with the rest of her morning's flier collection. She watched sadly as the papers flopped onto the overflowing pile. At least they were being recycled.

Three years ago, nobody even wanted to be a part of the student council. It had been just a dumb popularity contest with winners bossing around kids who generally ignored them. But that was before Benjamin Spencer.

Just an enthusiastic fifth grade council member at the time, Benjamin ascended to school president when the old one never showed up at meetings. Benjamin was the only one willing to step into the position. Since then, Wren had seen the school change.

Benjamin and his student council did actual things. Benjamin, now in eighth grade, led the school in an anti-bullying campaign. Benjamin convinced his aspiring journalist friend, Gail Mendez, to start a school paper, *The Lovelace Gazette*, that everyone adored. Benjamin, a swimmer, revived interest in school sports. Benjamin started a competitive math team and suddenly math trophies lined the front entry shelves, and math was cool. Benjamin, Benjamin, Benjamin. Even Wren was impressed by him, and she had pretty high standards for being impressed by anyone.

This year, two spots on the council had opened up and suddenly everybody wanted in. Kids were fighting for those seats like rabid social gladiators. Bright posters, covered in streamers and slowly deflating helium balloons, plastered every inch of the hallway with slogans in foil and fluorescent CAPITAL LETTERS. Their visual noise screamed mercilessly at Wren as she moved carefully down the hall.

Finally, exhausted, she reached her locker. It wasn't even nine a.m.

She smacked her head against its door a few times with a groan, then opened it. Hanging her backpack and jacket on the hooks, she pulled out her science book and turned to head to her favorite class.

Amber was waiting.

2

BE BORING, BE BORED

*a*mber Rosenberg's delicate, freckled face lit up like sunshine when Wren entered the classroom. Warmth spread through Wren as she sat at the science table next to her best friend. One of her best friends. She had three. Amber and the other members of the Renegade Girls Tinkering Club.

Mr. Malcolm didn't care where kids sat as long as they were paying attention, so Amber and Wren pretty much always sat together, and almost always paid attention.

As Wren sat down, Amber slipped the book she'd been reading, *Sarah and Simon: Super Spies,* into her backpack and leaned over excitedly, "Mr. Malcolm is setting up a hands-on experiment! He's so wonderful."

"If you say so," Wren laughed.

Whimsical, otherworldly Amber. Wren wasn't sure her best friend was entirely human. Tiny and clever, with wavy auburn hair, beautiful clothes, and graceful dance moves, Amber was fascinated by plants and animals. She wanted to know everything about all living things and couldn't be bothered to divide them into "cute" and "icky." She loved cockroaches as much as cats. Well, maybe not exactly as much, but she was a lot less judgmental than Wren. Secretly, Wren had decided that Amber was some sort of spy, studying Earth for her true people. Amber's species, probably fairies or aliens or something, would eventually use the information to become overlords of the Earth. Amber would rule them all. The thought was one of Wren's favorite daydreams but Amber knew nothing about it.

"What do you think those are?" Amber pointed towards some bins Mr. Malcolm was filling with small items.

Wren shrugged. She saw a baggie with some powder in it and a tiny bottle of liquid. A mystery!

Maybe they'd use the microscopes today. Those weren't boring, not even the ones at school, though Wren was kind of spoiled by the Renegade Girls's professional grade microscope. It had been a gift from Amber's Uncle Tim when his microbiology lab upgraded.

It was a thing of beauty. Looking through it transported Wren to another world--a microscopic world filled with fantasy and wonder. The millions of iridescent, feather-like scales on a butterfly wing, the bubbly geometric beauty of the cells of a plant stem, the shapes and colors in a pinch of beach sand. Intricate and beautiful structures made up everyday objects.

It was real life magic.

Amber felt the same way. That bond had blossomed into their friendship and then their club. It didn't matter to Wren if anyone else at school thought they were weird or nerdy. They had each other. Together, they understood the deep magic of the world.

Thinking about the microscope was definitely not boring, though. Wren reminded herself that her goal was to be boring. Be like the other kids who didn't care about important things, like microscopes, and never got in trouble. This was going to be harder than she'd anticipated.

"Okay, guys," Mr. Malcolm said. "We've been talking about states of matter. As I'm sure you all remember, matter is the stuff that makes up everything in the universe, everything that occupies space and has weight. Who can remind us what the basic states of matter are?"

Amber's hand shot up, along with most of the class.

"Emma?" Mr. Malcolm chose the dark haired girl to their right.

"Solids, liquids, and gases," Emma answered, punctuating "gases" with a fart noise. Wren rolled her eyes while the class giggled and Mr. Malcolm smiled tolerantly. "Which has something to do with, like, how many molecules whatever has, right?"

Wren's attempt at boredom cracked a little at the

mention of molecules. Maybe they really would use the microscopes! You have to use a microscope to see atoms and molecules, the tiny bits that matter is made of. Actually though, it had to be a special kind of microscope. Molecules were too small for even their club microscope to see.

"Exactly!" the science teacher nodded. "In solid things, the molecules are all crammed together too tightly to move around. They are fixed. You can have a 'chair shaped' thing or a 'squirrel shaped' thing. A chair and a squirrel are both solid. But in a liquid, the molecules are looser. They slide around. So a liquid doesn't have a fixed shape. Lemonade can be shaped like a cup, or a straw, or the inside of your mouth. There is no such thing as a 'lemonade shaped' object. You can't compress a liquid or a solid very easily. If you try to squish a mouthful of lemonade, the volume won't change, it'll just change shape and shoot right out of your mouth. But a gas is a different story. Molecules in a gas are really loose..."

A knock came from the door. Walking over, Mr. Malcolm said, "You're doing an experiment today. But first, I have a surprise. Mrs. Yang went into labor this morning right after she got to school."

Whispers surged through the class. Wren didn't think Amber could look more excited, but apparently tiny new babies broke all previous excitement levels.

"Well," he continued, "she's off to the hospital, and the

substitute hasn't gotten here yet. So this morning we're going to welcome her class for today's lesson. You'll have to double up your stations."

The classroom got crowded as more sixth graders streamed in. With a groan, Wren watched Axel Andrews and her perky blond ponytail saunter past, joining the table in front of them. Just the sight of Axel stressed out Wren. She desperately scanned the other newcomers, looking for friendly faces.

One had already appeared next to her. Silently, like a ninja.

3

SCIENCE!

*I*f "secret" were a state of matter, it would be the shape of Kaminia Doyle. Kammie's molecules of secretness were packed tightly, moving quietly and carefully. She rarely spoke to anyone besides her friends, so it was a secret that she actually spoke a bunch of languages. English of course, plus Hindi with her mom's parents and even some French. And to top it all off, she could talk to computers in different programming languages too. Not expert level or anything, but she was working on it. Yet even with so many ways to communicate, Kammie usually hung in the background, unnoticed. Most people thought she was boring, but that was her secret. Kaminia Doyle was one of the most fascinating people Wren had ever met, but you had to hunt for it. And Wren loved a good mystery.

Apparently, though, Wren had missed something in this class.

"Yes, Axel," Mr. Malcolm was saying. "That's very interesting. But the question was, does anyone know what a non-Newtonian fluid is?"

A few eyes skittered towards Amber and Mr. Malcolm pointed to her. "How about you give it a try, Amber?"

"That's a substance that can act like a solid or a liquid, depending on how you interact with it," Amber said confidently. "Like quicksand or ketchup."

Axel seemed jealous, "How is a squished up vegetable a non-Newton fluid? That doesn't even make sense!"

Amber reddened. "But it is! Ketchup is so viscous that it changes between solid and liquid depending on the forces affecting it. Really."

Axel stared at her. "Vixos isn't even a word, Amber."

Amber tilted her head in confusion as Mr. Malcolm watched the two girls with interest. "Wait, what?"

Axel, encouraged by Amber's confusion, opened her mouth to gloat.

But Amber wasn't finished. "Don't you know what viscosity is? It's the rate a liquid flows. You know, like how thick it is. And how fast ketchup flows changes based on outside forces, like when you hit the bottle. That's what makes it non-Newtonian. And by the way, a tomato isn't even a vegetable. It's a fruit."

"BAM!" Wren blurted. "You just got SCIENCED!"

Mr. Malcolm stifled a laugh.

"That's not really the cool comeback you think it is," Axel mumbled.

This class was totally not boring so far. Wren tried to stuff down her emotions by copying Bobby's tuned-out expression at the next table. She wished Ivy were there. Ivy Rose Park, the fourth member of the Renegades, was good at being businesslike and serious. But her science class was in the afternoon.

"Alright, everybody! Back on track," Mr. Malcolm passed out a bin to each table. "Today you'll be making... wait for it... SLIME! And Oobleck too. Work in groups. You'll find recipes and instructions in your bins. Oobleck and slime are both non-Newtonian fluids which, as Amber pointed out, are neither solids nor liquids, yet have characteristics of both. Non-Newtonian fluids are a fourth state of matter."

"What about plasma?" Wren asked as he set down their bin. "Isn't that another other state of matter?"

Mr. Malcolm paused and gave her a curious look. "Indeed it is, but that's a bit beyond the scope of today's class. Where did you learn about plasma?"

"YouTube," Wren replied absently, already digging in the bin with Kammie.

"Apparently, not everyone uses YouTube the same way," Mr. Malcolm chuckled. He continued on. "Follow the directions in your bins. The two substances are simple to make, but messy. Work INSIDE your bins. Take out your science notebooks and record your observations. Compare and contrast both substances. Everybody got it?"

Everybody got to work.

4

SLIME!

$\mathcal{W}$ren and Kammie started taking the contents out of the bin. Kammie stacked them in neat little piles with a matching instruction card in front of each. She was cross referencing the ingredients as Amber picked up a card.

"He's premixed the borax water for the slime," said Amber, mostly to herself. "One and a half teaspoons borax to four cups of hot water. That's pretty much what I use at home."

"Wait!" Kammie squeaked. "Isn't borax toxic?"

"Nah," Wren replied. "It's an irritant, but only dangerous if you use a whole lot of it. Mom looked it up. She even printed it out and put it on the fridge with all that science and engineering stuff she's always sticking up there. Did you know the reason you can use contact lens solution to make slime is because it has borax in it? Like you put that stuff in your eyeballs!"

Amber looked up in alarm, "Don't put this in your eyeballs, Wren! That's a terrible idea. And we should wash our hands afterwards. With soap."

"Never put anything in your eyeballs, ever, not even as a joke," Kammie agreed as Amber handed the card to Wren. "So what do the slime instructions say?"

"Who cares?" Wren tossed the recipe back into the bin. "Lets get messy!"

"No!" Kammie smacked Wren's hand as she reached to grab a bottle. "This is serious science! We follow the rules."

Wren pouted.

Kammie ignored her and carefully poured a half cup of the boraxy water into the mixing bowl.

Amber measured out an equal amount of glue, then dripped a little purple food coloring into it and stirred slowly with a jumbo craft stick from the box. All three girls paused to watch as rich purple color swirled through the thick white glue. Soon all the glue was evenly purple.

Then it was Wren's turn to stir while Amber poured the glue into the borax water and Kammie held the bowl. Soon the glue coalesced into a blob. Wren grabbed the sticky, gooey glob and began to squish it with her hands. Again and again she squeezed her hand into a fist, letting the substance squirt thickly through her fingers and then catching the gooey ribbons and folding them back in. As she kneaded, it got more and more solid. She had to dunk it in the leftover borax water a few times to firm it up, but soon had a nice ball of purple slime.

Meanwhile, Kammie and Amber had taken out the second mixing bowl and a baggie full of white powder. It was two cups of corn starch. They dumped it in the bowl and added one cup of regular water. As they slowly stirred, the mixture started to look kind of like yogurt. Watching them, Wren wondered if yogurt were a non-Newtonian fluid too.

She held up her slime and let it sag and drip thickly over the edges of her hand, looking at the difference in how the oobleck ran quickly through Amber's fingers like a liquid. The slime didn't ever get that runny. It was pretty much an oozing, slow moving mass. The oobleck was a LOT more liquidy than the slime when Amber left it alone, and almost solid when she smacked it. Wren imagined she could run over a pool of oobleck if her feet slapped it hard enough, but probably not slime. Apparently even two non-Newtonian fluids could be different. Kammie wrote notes with her quick, tidy handwriting.

Aaaaand Wren was getting bored. Bored was what she'd promised. Bored was going to keep her out of trouble, right? So why did it feel so dangerous? Setting the slime in Kammie's outstretched hands, Wren wiped her own off with a paper towel.

Then it happened.

As she set the wadded paper towel on the table next to a few craft sticks, a rubber band, and the little empty plastic measuring cup, pictures popped into her brain about ways they could all fit together.

A stretched rubber band snapped back with serious force. If two stiff objects, like the craft sticks were rubber banded on one end with something in the middle, would it act like a teeter-totter? The rubber band would stretch when it teetered, and snap back when it tottered. Would it snap back with a lot of force like a rocket-powered teeter-totter?

Or a catapult.

And suddenly all the pieces formed together in her head. Her hands started to assemble before she knew what was going on.

She laid two jumbo craft sticks on top of each other, wrapping the rubber band around one end to hold them in place. Then she wedged the wad of paper towel between them, pushing them apart into a V. She set it flat on the table and, bracing the bottom craft stick with one hand, pulled down on the end of the top craft stick. To her surprise the rubber band didn't really stretch like she thought. Instead, the stick curved. When she let go, the stick shot back straight. Not what she'd planned, but maybe it would work anyway.

Almost in a trance, she taped the tiny mixing cup to the end and dropped in a piece of rolled up tape. Holding the bottom stick again, she pulled and released the cup end. The tape went flying. Catapult!

Then Wren looked at the slow moving slime in Kammie's hands. It moved like thick mud. But the oobleck's non-Newtonian state was a lot more liquidy unless something hit it hard.

Or it hit something.

Wren didn't notice Amber and Kammie look over at her in surprise as she poured a big glob of the oobleck into the catapult's cup. Would the oobleck stay liquid in the air? Would it turn solid when it hit something? She couldn't hear Mr. Malcolm saying something sharply at her or see him moving quickly towards her, reaching out.

She pushed the paper towel closer to the rubber band, increasing the angle of the upper craft stick to give it more power, bent it down, and shot it, watching with fascination.

The oobleck didn't fly out like she'd thought it would. Instead, the force of the motion turned it solid enough that it clung to the little plastic cup. The weight of the cup and substance was too much for the small piece of tape holding it in place though, and the whole cup broke free from the device. Too late, she realized she had no idea where it would land.

Plastic cup and non-Newtonian fluid soared together in a graceful arc across the short distance to the table in front of them, rotating slowly along the trajectory.

Without a sound, the tiny plastic cup plopped right on

top of the perky blond head of Axel Andrews, open end down like a little hat.

Harmlessly, the cup fell to the floor. But its contents remained glopped on top of Axel's head. Then it turned back into a liquid and slowly oozed down her head, through her hair, and into the inside of her shirt.

Axel let out a high-pitched howl, snapping Wren out of her trance. Mr. Malcolm and his angry face came into focus. She could hear laughter from the other kids in class. Wren, suddenly realizing what she'd done, looked over at Amber and Kammie who were frozen in surprise, staring at her.

The whole class erupted in chaos. Mr. Malcolm, furious, pointed towards the door, sending her to Ms. Sophie's office. The office she'd just left.

She'd already broken her promise.

Wren realized at that moment that she would never be able to do it. She'd always be a problem starter. A trouble-maker. Everywhere except the Renegade Girls Tinkering Club. It wasn't even worth fighting anymore.

CATAPULT!

MATERIALS:
1 Wad of Paper
1 Rubber Band
2 Jumbo Craft Sticks
1 Sauce or Dixie Cup
Some Tape

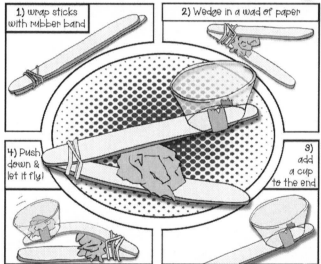

1) wrap sticks with rubber band

2) Wedge in a wad of paper

3) add a cup to the end

4) Push down & let it fly!

NOW MAKE IT BETTER!

What flies better, heavy stuff or light stuff?

What else can you change?

Where else could you put the cup?

What could you use instead of paper? Instead of a cup?

5

THE GREENHOUSE

*T*echnically it was Wren's Greenhouse, hidden behind her family's small home in the middle of San Francisco, but all the Renegades felt at home there. In the Greenhouse, they didn't have to worry about other people's rules and opinions. Or try to be boring or be like everyone else. The Greenhouse was their safe space. They could just be themselves. Wren's parents let them use it as their workshop and clubhouse as long as her little sister, Trixie, could be part of the club.

When they'd formed the Renegade Girls Tinkering Club, the Greenhouse had been abandoned and filled with broken pots and spiderwebs. It was small and dirty, but had everything they needed. A door hidden like a secret behind an overgrown wisteria vine. A back wall with shelves from floor to ceiling, and excellent light from a front wall made entirely of glass. A small but sturdy potting table sat against the windows. It was pleasant and warm, with one electrical outlet and a small work sink. They loved it from the first time they saw it.

Amber, Kammie, Ivy, Wren, and even Trixie had worked tirelessly last summer, cleaning and gathering assorted leftovers, recyclables, and a mishmash of bins to put them in. They categorized and labelled, collected cardboard by cutting down shipping boxes, and saved empty toilet paper rolls from the trash. They snuck random scissors from kitchen drawers, ribbons, buttons, anything that looked useful or had an interesting shape. Amber had borrowed a folding card table from her garage, and Kammie brought in some stools her parents were getting rid of. Wren found an old glue gun, and they had even managed to find an unused sewing machine. The first purchase with their club dues had been copies of the side gate key, so everyone could head directly into the backyard when they came over.

Amber rocketed through that side gate, clutching a cardboard box protectively to her chest with her delicate arms. Beneath a spring green sundress her feet, in their pristine white flats, skipped quickly and skillfully over the ground. The September afternoon sun lit up her auburn hair like a fiery halo.

Amber shoved her way through the overgrown wisteria near the door, but Kammie had gotten there first. The quiet, dark-haired girl had seen Amber coming, and was carefully placing her own copy of *Sarah and Simon: Super Spies* into her backpack.

"Is that what I think it is?" Kammie asked, indicating the box as Amber placed it on the potting table.

"They finally came!" Amber sang, her hazel eyes sparkling with excitement. "Where's Wren?"

"Looks like she's on her way." Kammie pointed at Wren's house.

Wren strode towards them across the yard, wiping some lunch crumbs from her stained cotton dress. A breeze ruffled her unruly mop of light brown hair.

"I'm glad her mom didn't ground her for that thing with Axel yesterday," Amber confided. "I was afraid we wouldn't be able to meet."

Kammie nodded without saying anything and waved at Wren through the window.

In the background, Trixie's little face and hands pressed against the sliding glass door from inside the house, watching as Wren walked away. Trixie was almost six years old and generally sticky. Sure enough, as she peeled off the back door to return to the table, prints from her lunch remained on the glass.

"Is Ivy coming?" Wren asked, closing the Greenhouse door behind her.

"I called her this morning," said Amber. "Her mom said she'd drop her off right after the soccer game."

"Oh good," said Wren. "Aren't they playing McKinley today? I hope she won."

Amber didn't reply as she carefully dug into a bin marked SHARP STUFF and pulled out some scissors. Apparently Wren didn't want to talk about the catapult, and Amber certainly wasn't going to be the one to bring it up.

Holding the scissors over the package with a dramatic gesture Amber sang, "What will we fiiiind inside the boooox??? It's a mystereeee for yoooou and meeeeeee!"

Wren wriggled on her stool. "We know what's in the box! Open it!"

Amber opened it. A smaller wooden box huddled inside. She pulled it out and lifted the lid. Row after row of glass rectangles glittered in slotted holders. It had taken the Renegades four weeks to save up for those microscope slides. Seventy prepared slides and thirty blanks.

The mysteries of the universe awaited on those little pieces of glass.

Wren moved their club microscope into place. It was heavy, solid, and powerful, but she still handled it like it was made out of newborn puppies. She gently pulled off the cover and plugged it in. It was beautiful. To Amber's Uncle Tim and his lab, it may have been obsolete tech but to the Renegades it was the most professional piece of science any of them had ever gotten their hands on. And it was all theirs. It had two sets of binocular lenses on the viewing head, the type of eyepieces you could look through with both eyes at once. Between them and the three rotating lower lenses, it could magnify up to two thousand times. It had back lights, front lights, and even a camera attachment.

The ten slides Uncle Tim had given them had been explored, researched, compared, and contrasted to death. Wren could draw the cells of a papyrus stem (slide #3) by heart, and she wasn't entirely sure what papyrus even was. So they had decided in the middle of the summer that it was time to purchase more slides. Even using their whole treasury, it had taken them weeks of extra odd jobs and allowances to earn enough money for the new slides. And finally, here they were.

The girls heard the creak and slam of the side gate as Ivy careened into the backyard. She tossed her soccer bag next to the Greenhouse door before bursting in breathlessly. Taller than her friends, she peered right over Wren's head into the box.

"Awwwwww, you opened it without me?" Ivy complained, redoing the ponytail in her long black hair. She smelled like sweat after running around on the field for ninety minutes in the sun.

Kammie slid their inventing journals across to each of them. A fresh section of butcher paper already sat on Trixie's old art easel, also known as their noteboard. Kammie pointed to it. She had started a "Microscope Viewing Order" list and in small, almost apologetic lettering, her own name was in the top spot.

"I mean, I WAS here first," she blushed. "So I thought, you know, maybe this time, if you guys don't mind..."

Amber snagged second place and Ivy slotted in at third.

One of the oldest kids in their grade, Ivy Rose Park was not used to being second, let alone third. Tall, confident, and always getting picked for some sports team or another, Ivy was not friends with the concept of waiting. Wren didn't like the word "bossy," but she'd heard it a lot when people talked about Ivy. Wren could see from her friend's pursed lips and how she kept stopping herself from tapping her pencil that Ivy was trying to wait patiently. She'd agreed to the order and once Ivy agreed to something, she stuck with it like glue. Or at least she did her best to.

Kammie was not helping by methodically, slooooowly trying to pick EXACTLY the right slide. She didn't get to go first very often and didn't want to waste the opportunity. Plus, she was probably afraid to get it wrong. It could be a real problem, Kammie's fear of messing up. She hated being wrong so much. Wren didn't understand. She, herself, was often wrong. She'd learned to roll with it because she had to. But when something didn't work out for Kammie, she completely crumpled. More than once Wren had sat and rubbed Kammie's back as the quiet girl sobbed and scolded herself for tiny little mistakes. Because no matter how many times Kammie said she was stupid, she wasn't, and Wren could never figure out why she thought she was.

Ivy, trying not to push, watched Kammie walk her fingers over the slides again, reading and evaluating each label. As Kammie started back at the beginning of the row, Ivy groaned and reached into her backpack. She pulled out her latest book on Arduino, or the Raspberry Pi, or whatever electronic doodad she was obsessed with this week. Ivy wanted to be an electrical engineer like her mom, so she spent what free time she had outside of Renegades, school, and her various sports learning about technology. Wren couldn't understand half of what she was saying sometimes, but the stuff she made usually did what it was supposed to do. Which was helpful when they wanted to attach a light or a buzzer to one of their projects.

Wren turned back to Kammie and started to say something to nudge her along but Amber beat her to it. She leaned past Kammie and plucked a slide out of the box at random, startling the serious expression right off her friend's face.

"Plant tissue! Fascinating..." Amber brought her slide to the microscope immediately, hopping to the front of the line. Kammie, with a panicked look, frantically pulled out the closest slide.

Frowning as she read the name Carpet Fibers, she handed the box to Ivy, who quickly decided on a slide with cat whiskers. Kammie pouted but waited patiently for her turn. Second after all.

Wren pondered the slides. She wished she had some of the slime or oobleck to dribble onto a blank, but she wasn't allowed to touch any more non-Newtonian fluids for at least a week.

"Wren! You have to see this!" Ivy called, pointing to the cat whisker slide that was now mounted on the microscope tray. "I know how much you love cats!"

She put her eyes to the eyepieces, but only glanced at the whisker. At five hundred times magnification, she could see the layered lines of scales along the shaft of the whisker. It was fascinating, but it was too late. Her mind already had different plans.

"What did you pick, Wren?" Amber asked.

"I picked this one," Wren held up a blank slide and winked.

"Yes!" Amber squealed. "Let's make our own!"

Kammie raised her hand. "We should set a limit on how many we can make though. We don't want to use them all up too fast!"

"Good idea," Ivy agreed. "One a week each, max?"

Kammie did a quick calculation, "That should last us about a month and a half..."

They were already collecting specimens when Trixie wandered into the Greenhouse.

THE MICROSCOPE

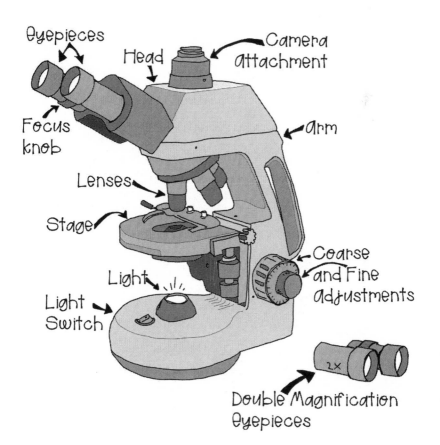

Eyepieces

Head

Camera attachment

Focus knob

arm

Lenses

Stage

Coarse and Fine adjustments

Light

Light Switch

2x

Double Magnification Eyepieces

6

TRAGEDY

"*I* washed my hands!" Trixie declared proudly, holding them up for inspection. "Are those the slices for the telescope? Are we making our own?"

"Microscope, yes. We each get one a week. Understand?" Amber explained, as gooey green slime from a succulent leaf ran through her fingers and over her pink painted nails. It dripped thickly onto a waiting blank slide.

"Okee dokee!" agreed Trixie, and approached the box.

Everyone worked silently for a few minutes.

Suddenly, Wren looked up from the pipe cleaner she was de-fluffing onto her slide. "TRIXIE! What are you doing?!"

Trixie, her tongue on a slide, looked up in surprise.

"Wat?" She removed her tongue. "I'm checking for little bugs in my spit, of course."

Wren's voice got louder. "I mean why do you have TWO slides?!?"

Trixie raised her left hand, holding the slide she had just licked. Then she raised her right hand, holding a slide she had dribbled spit on.

"This one is spit spit and this one is licked spit," she declared proudly, holding both slides up at once. "I made them faster than you guys!"

"That's not fair!" yelled Wren. "You're using them all up! You're only supposed to have ONE slide! You're cheating!"

"I AM NOT CHEATING!" Trixie howled back. "THIS is one, and THIS is one. I have TWO ones!"

"Guys," Ivy said, "It's really not a big deal."

It sure felt like a big deal to Wren.

"You don't get two ones, dumbface!" she insisted. "Give me those!"

Wren grabbed for Trixie's slides. Trixie had already hidden them low behind her back.

Immediately weaponizing her tongue, Trixie tried to lick Wren's reaching arm. Wren growled and grabbed her sister's head in both hands, holding it still.

"Somebody get those slides!" she yelled as her hip bumped into the potting table.

Amber reached over with one hand from where she sat at the potting table. But Trixie, slippery as an eel, shifted her body at the last minute, twisting her head free and trailing her tongue angrily against Wren's now empty hands.

"Ewwww!" squealed Wren, wrenching her hands back as Amber crashed to the floor with a yelp, one leg thrown ungracefully over the stool that had toppled over with her. A clean white shoe shone in the sun above the table.

Ivy stood and held up her hands in what she hoped was a calming gesture. "Now wait, let's just..." she began. Clearly no one was listening, so she shifted tactics and made a lunge for the now-feral kindergartener. Unfortunately she didn't have much experience with siblings, and Trixie easily avoided her.

Amber, who had two brothers, was already up, uninjured but pissed off. She straightened her white cardigan while tracking Trixie's erratic movements. Amber's hand darted out to grab the younger girl's arm. She caught it, but Trixie had had a chance to refuel her tongue. She licked her with a fresh supply of spit.

Amber quickly let go.

Both Kammie's hands flew to her mouth, her eyes huge as she pressed herself back against the wall of the Greenhouse. "Calm down, everybody," she whimpered, too quietly for anyone to hear.

Wren, blue eyes blazing, grabbed Trixie forcefully from behind. The younger sister, a slide still tightly clutched in each fist, wriggled furiously.

Suddenly, one of Trixie's hands smashed against the card table.

The sound of glass cracking and a cry of pain was quickly followed by the tick tocking of the microscope teetering on the edge of the table. It was so quiet they shouldn't have been able to hear it, but they did.

The Renegades froze in one giant mass, all eyes watching in horror as their microscope, their beautiful, irreplaceable microscope, slid heavily off the table and landed on the nearby stool with a solid thunk. In slow motion, the set of double magnification eyepieces slid off their mount and crashed to the floor with a sickening crunch.

A piercing wail came from Trixie, who held a glass slide that had snapped in her fist. Shaking, she opened her fingers and the two pieces of glass slid away. Blood oozed up from her palm and between hysterical cries, she looked like she might faint.

7

SHATTERED

Wren stared at her sister as Trixie broke into long, trembling sobs. Amber and Kammie were frozen in place, staring at the microscope.

Ivy recovered first and gently lifted the microscope back onto the table. She pushed it to the middle where it wouldn't fall again. With a shaking hand, Ivy reached for the set of eyepieces on the floor. As she lifted them, they all heard the faint tinkling of glass from inside. Ivy and Amber shared a worried look.

Meanwhile, Trixie wasn't getting any better.

Wren couldn't process what was going on. It was her fault. Here. Here at the Greenhouse, the one place she could relax. The one place she didn't mess up all the time. Wren's brain filled with fog. If she was a troublemaker here in the Greenhouse, was there anywhere left in the world she could be good?

Kammie was checking out the microscope, flipping the various lights on and off, looking through the other set of eyepieces and clicking the lenses into place one at a time.

Ivy held the broken eyepieces up to the sunlight and tried to see inside. "I think these are shattered."

Shattered.

The word hit Wren like a fist. Shattered. Because of her. Why had she made such a big deal out of everything? No one else seemed to mind what Trixie was doing. Why couldn't she be like everyone else? Why couldn't she be boring? Panic crept slowly up her throat. Without the microscope, would the others even come to her house? Why would they hang out with a troublemaker?

Trixie continued to cry. One looked at her squinched up little face and Wren shook herself into action. At least she could help her sister.

Taking Trixie by the shoulders, Wren took a deep breath and said, "I'll be right back, guys." She steered Trixie outside without looking at the wreckage of their equipment. Hoping they'd still be there when she got back.

Their mom looked up as Wren led Trixie into the kitchen through the patio door.

Immediately dropping the knife she'd been chopping with, their mom ran to Trxie, wrapping her in her arms.

"What on earth, Wren?" their mom said. "What happened?"

Wren cringed. "There was an argument and she broke a glass slide in her hand. It got cut."

Trixie gulped a few times and opened her palm to show her mom the blood.

Wetting a paper towel at the sink, their mom addressed Wren angrily, "For heaven's sake, Wren, what have I told you about your temper? I'm getting really tired of it. Sometimes you only think about yourself..."

Turning, their mom saw Wren's face for the first time. She stopped in her tracks.

"I know," Wren whispered to her feet, her shoulders sagging. She wiped angrily at her face with the back of her hand.

"Oh honey," her mom said, kneeling down to blot at Trixie's hand. "I'm sorry. Are you okay? What's going on with you?"

"I just keep messing up. I can't do anything right," Wren watched as her mom pushed away Trixie's floppy hair to kiss her forehead. "I'm just a troublemaker. Why can't I be like everyone else?"

Wren's mom looked at her, "Honey, you were never meant to be like everyone else. You're exactly who you should be."

"A troublemaker?" Wren's bottom lip quivered.

Her mom smiled, "Sometimes. Yeah. But trouble is just another word for adventure. Excitement. It's all in how you look at it. And what you do with it."

Wren sniffled and smiled a little.

Trixie had settled down and was nestling against their mom, who picked her up and started to carry her towards the bathroom. Suddenly, their mom stopped and turned back to Wren. She gently set Trixie down and placed her hands on either side of Wren's face, looking deep into her big blue eyes. Wren fidgeted uncomfortably and tried to look away but her mom held on.

"You've got this honey. I don't know what trouble you're talking about and I have to take care of your sister right now, but whatever the problem is, I know you can handle it. You can do hard stuff. You're stronger than anyone I know. Just, I don't know, start with some structure. Plan a little bit first. I believe in you."

Wren watched her pick Trixie back up and walk away down the hall. She looked around the kitchen. Well, at least maybe she could bring snacks. She stuck some popcorn in the microwave and grabbed the lemonade from the fridge, along with some plastic cups. The microwave beeped.

As she pulled the hot popcorn out and dumped it into a bowl, she noticed a piece of paper taped to the fridge among the mess of printouts her mom kept pinning up there. It said "Engineering Design Process." The word Process was what caught her attention. She vaguely remembered her mom saying it was a way to solve problems. And she could certainly use help solving problems.

She grabbed the paper off the fridge with a shrug and headed back to the Greenhouse.

Maybe there was still a chance to fix things.

8

THE ENGINEERING DESIGN PROCESS

"*T*rixie's in the recovery ward," Wren backed into the Greenhouse with arms full of popcorn and drinks. The paper from the fridge dangled from her hand. "How's our other patient?"

"Not good." Kammie replied softly, still checking the microscope. "But it could have been a lot worse. The scope itself seems okay. So do the regular eyepieces. But I think the double magnifying eyepieces are broken."

Ivy held up the set of eyepieces before shoving them and the box of slides onto a packed shelf. She gently unplugged the microscope and pulled the cover over it like a shroud, and carefully put that away too.

"I know this is all my fault," Wren dropped her snack offerings on the table and flopped onto a stool, unable to meet their eyes. "But I promise I'll do whatever it takes to fix things."

Ivy looked up from the shelves, and Kammie and Amber looked at each other in confusion.

"What do you mean?" asked Ivy. "It was an accident. It's nobody's fault."

"It IS my fault," Wren insisted. "I just can't figure out how to be boring! No matter how hard I try."

"What? Wren!" Amber grabbed her friend's shoulders and shook her gently. "Of course you can't be boring. That's one of the best things about you!"

"You don't understand. I'm a troublemaker."

"Cut it out, Wren," Ivy crossed her arms. "Everybody was involved here, we'll fix it together. It's not the end of the world. The only question is what are we going to do about it?"

Wren blinked. They weren't going to yell at her? Not only that, they didn't even seem mad. She'd really lucked out in the friend department. She wanted to hug them.

"Thank you for giving me another chance, guys. You're really awesome."

"Um, thanks, I guess," Kammie shrugged. "But let's get back to solving this. Because I don't even know where to start."

"This might help." Wren held up the paper from the fridge. "It's the engineering design process. Mom says engineers use it to solve big problems. It's sort of like a step-by-step way to do stuff. I think. I wasn't really listening to her, but I thought we might give it a try."

"I love processes and systems!" Kammie squealed excitedly. "What does it say?"

Wren taped the paper to the wall next to the easel and pointed to the first gear. "What's the problem?"

"I don't have a problem," Kammie frowned. "It was just a question!"

"No, no," Wren replied. "I mean that's the first step. Figure out the problem you're trying to solve."

"Well that's easy." Amber said. "Our problem is the microscope broke."

But Wren shook her head. "Actually no, I mean, that's not a problem, right?"

"Not a PROBLEM?!" blurted Amber.

"Not really. It's a fact. And only the eyepieces are broken. So the problem is how do we get new eyepieces!" Wren replied.

"Our parents will buy them," Amber said. "Problem solved."

"Except," Kammie raised her hand, "everybody said if we were going to have an expensive microscope out here, we had to take care of it. If we tell them we broke it and ask for money, I'm afraid they'll take it away."

"Yeah, we need to solve this ourselves," Ivy agreed. "And we don't have enough in the treasury after buying the slides."

"So then," Wren pondered, "the real problem is how do we get new eyepieces with no money."

They all nodded.

"See?" Kammie beamed. "Systems work! What's next?"

Footsteps came from outside and Trixie opened the Greenhouse door, her hand wrapped in way too much bandage.

"Can I come in?" she asked as she came in. "It wasn't my fault."

Wren smiled and patted the stool next to her.

"What's that for?" Trixie eyed it nervously.

"It's for your butt," Wren explained. "Sit down and help us solve our problem."

"What problem?" Trixie inched the stool away from Wren and sat down.

"We need a way to make money to buy new eyepieces," Wren yanked the stool closer, with Trixie still sitting on it.

"How much money?" Trixie asked. "I have three dollars."

"Aww thanks," Amber smiled. "But that's not going to be enough."

"Three hundred dollars?" Trixie asked. "Or three thousand?"

Amber shrugged. "Let's just go with 'a lot.'"

Wren pointed at the paper. "Next step is to Imagine, or Brainstorm."

"Hey!" Trixie exclaimed. "That's from the fridge!"

"Don't tell Mom I took it," Wren whispered.

"Brain storm?" Trixie tilted her head, "Do we need brain umbrellas?"

"No," Wren replied. "It means thinking up a bunch of ideas at once and then picking the ones you like best. We do it at school sometimes"

"It's actually really helpful," Kammie uncapped a marker and brought it to the easel. "Wren's great at it. I'm not very creative but even I've had a few good ideas. My favorite is writing it all down, though."

"Whatever. I'm glad you're the one writing. My handwriting sucks," Amber shrugged. " So what can we do to make money?"

Trixie, already bored, pulled Wren's copy of *Sarah and Simon: Super Spies* out from under a pile of cardboard and began flipping through it looking for pictures.

"What about a car wash?" Ivy suggested. Kammie wrote it down.

Wren grabbed a handful of popcorn. Noticing Trixie had her book, she yanked it out of her hands and set it on the other side of the table. Trixie frowned and grabbed some popcorn too.

"Oh wait!" Kammie pointed to Wren's face as popcorn crumbs spilled from her mouth. "Everyone likes snacks. What about a bake sale? I love baking, and Amber makes fantastic cupcakes."

"Yay!" cried Trixie as Kammie added it to the list. "Brownies! And carrot muffins!"

"There's only one problem," Wren swallowed the last of her popcorn with a sip of lemonade. "With the elections, everyone's handing out free snacks. I almost got a faceful of brownie Thursday when Andrea Patinkin caught me by surprise. I'm totally not voting for her."

Amber sighed. "Yeah, they're everywhere. And Axel's lemon cupcakes really are fantastic. Mom was mad I didn't finish my lunch on Friday but, those cupcakes... I might have accidentally had two."

"Seriously, no one's going to vote for those guys," frowned Kammie. "They just take the food."

"Vote or not, Wren's right." Ivy agreed. "They won't buy our stuff if they can get it for free."

"Okay, good point, so other ideas?" asked Kammie.

"What does the Renegade Girls Tinkering Club do best?" Wren picked up her *Sarah and Simon* book and tapped the table with it absentmindedly.

"Tinker?" offered Trixie tentatively.

"Yes! We tinker! We make inventions and use our glorious brains to build stuff, right?" Wren spread out her arms like she was hugging the entire Greenhouse. "We persist, we solve problems. If we want to raise money, we just have to make inventions to sell!"

Ivy perked up, "That's true! School doesn't have an inventing club. We'd have no competition!"

Amber leaned forward. "That's actually a really good idea." Trixie beamed up at her proudly.

Kammie started to write down "sell inventions," but turned back to them after the "t." She looked skeptical. "Wait, are you talking about doing a real business-type thing?"

"Sure, I guess," Wren said indifferently. She started drawing spirals on the inside cover of her book. "Why not? What could possibly go wrong?"

Amber scowled. "You did NOT just say 'what could go wrong'! That's like asking things to go wrong!"

"It's kind of a bigger deal to sell inventions than bake cupcakes," Kammie stammered. "We have to figure out what to make and how to make it and buy materials. What if we need special tools? What if kids don't buy anything? What do we do with our surplus? And what if an invention breaks? Is the kid we sold it to going to ask for their money back? Do we need tech support for our inventions? Plus, figuring out how much to charge for each item is more complicated than it seems..."

"Oh blah, blah," Wren interrupted, then quickly softened her voice when she saw Kammie's stricken face. "Sorry, I just mean, you know, everything has its problems. Aren't we problem solvers?"

"I'm sure we can work all that out." Amber said soothingly. "Right now we're just brainstorming."

"I dunno, I think we found our winner," Wren replied. "What we should be brainstorming is WHAT we're going to invent and sell."

Ivy nodded in agreement.

"If we're starting a business," Kammie relented, pulling up fresh paper on the easel, "we have to be really logical about it. We need to figure out what kind of inventions to make, keep track of our expenditures and our inventory. Then by comparing the cost of construction with the market value of each item, we..."

A low groan came from Wren as she slouched on her stool, tilting her head so far back she almost tipped over. "This is taking sooooo loooooong."

Kammie pursed her lips together angrily. "Look Wren, I think this business is going to be trouble. We need a plan or it's not going to work. Seriously."

Wren sighed noisily, drawing her copy of *Sarah and Simon* closer, flipping the pages mindlessly. "I'm just getting bored. BORED. I want to start making stuff now."

Amber smirked, "I thought you wanted to be bored, Wren."

Wren paused. Did she? Was that really the answer? "Well, maybe not bored. But if I can be boring, I'll cause less problems."

Amber shook her head, "I don't think it works that way."

"I promised Principal Sophie I'd be boring, I'd be like everyone else," Wren insisted.

Amber frowned. "We're boring?"

"No! No!" Wren felt trapped. "That's not what I mean! I mean GOOD boring! Sorry, that came out all wrong."

"Hmph," Amber hmphed. "Whatever."

Ivy clapped her hands. "Back to work!"

"I still think planning is a waste of time when we can be building," Wren mumbled, trying to look inside her book without anyone noticing.

Even though they were all looking right at her.

Kammie watched Wren open the pages a little more. "Okay, tell me this. What would Sarah and Simon do before starting a spy mission?"

Wren looked up. "They'd make a plan. With lists. And do field training with Agent Zero."

Sarah and Simon, the teen-aged spy twins, routinely saved the world by blowing up Doctor Dinklemeyer's evil inventions and returning brainwashed gorillas to the London Zoo. Agent Zero always loaded them up with gadgets and training before sending them out to save the world. Wren could see the importance of a good plan when Kammie put it that way. Without Agent Zero's tireless efforts, the world would probably already have blown up twenty times over. Even Sarah and Simon couldn't tackle a mission without proper preparation.

Besides, the engineering design process had gotten them this far. Maybe Kammie was right. Maybe, instead of being boring, she just needed a plan. Plans did seem like a good way to solve problems, even if they felt like a waste of time.

"Okay, let's start from the top," Wren threw herself into it. "What's the Problem? In this case the problem is what inventions we should make. What do kids want to buy?"

"I want to look at it," Trixie grabbed at Wren's book.

"Wait your turn," Wren held it up out of her reach.

"Awww. I love spy stuff," Trixie pouted.

"That's it!" Wren looked at Trixie in amazement. "Spy stuff! Everyone loves spy stuff! We can make spy gear inventions! The kids will pay bazillions of gajillions for anything spy related. Plus it would be FUN to make!"

"GREAT IDEA!" Amber gushed. "Spies are so cool! I love it when Sarah and Simon go undercover at fancy balls, like that time they visited Doctor Dinklemeyer's lair for his winter gala. It all sounded so sophisticated. Hey, has anybody figured out if Doctor D is really a spy too?"

"I think he's really The Shadow," said Wren.

"Isn't The Shadow a secret agent? Are spies and secret agents the same thing?" Amber asked. "Doesn't The Shadow work for the government?"

"What makes someone a spy?" Kammie asked.

"I think spies and secret agents are the same thing," answered Ivy. "They both use codes and sneak around."

"Yeah, spies are super sneaky and get into a lot of trouble," Wren offered. "Not unlike me."

"Sarah and Simon are spies!" Trixie added. "They

rescued Bobo the brainwashed gorilla, and found the Sapphire of the Panda King"

"If spy stuff is stuff a spy uses," Wren pondered, "maybe we can figure out what stuff to make by thinking about what they need to do. Like, what do you need to be sneaky? Maybe Shoes of Silence. Or maybe something so you can see from far away, or around a corner so you don't even have to be quiet. Shoes of Silence and a Corner Looker-arounder would be spy gear. What else?"

"Maybe codes? Something to hypnotize a gorilla? How about a radio to talk to your partner?" suggested Amber.

"OH! Sarah used a grap-a-link hook to climb the Tower of Terror! That's spy stuff, right?" asked Trixie.

Kammie wrote PROJECTS on the easel. Underneath, she listed every idea she heard, including the crazy ones, as they started calling them out.

"Magnifying glasses find hidden stuff..."

"Fingerprints tell you who was in a room..."

"Spies break into safes! Safes are good for keeping secrets."

"Suction cups to climb up the outside of buildings!"

"Simon once had a snorkel kit hidden inside a tube of toothpaste."

"Lollipops!" added Trixie insistently.

Kammie wrote down "fingerprints" too, even though they weren't a gadget. She could change or cross it out later but while the ideas were flowing, she just wanted to get everything written down. By the time they ran out of ideas, there were twenty-two items on the list.

"Wow, that's a lot of possibilities!" Kammie declared, massaging her hand.

Trixie looked terrified. "But... do we have to make ALL of them?"

"Nah," Ivy reassured her. "We can cut down the list by thinking about what we can do, like what we have time and materials for."

"A periscope, like on a submarine, lets you see the outside from the inside. What about something like that?" Wren suggested, pointing to "a corner look-arounder."

"Good idea," Kammie put a single line through it and wrote "periscope" instead.

"Put a star by that one," Ivy suggested.

"And codes," added Wren. "Lots of ways to make code gadgets and decoders cheap, easy to carry, and fun. Star that one too!"

"What about the camera necklace? It would be so pretty if it's like an aqua colored jewel on a gold chain, and kids would totally buy it," said Amber dreamily.

Kammie looked doubtful. "I don't think it exactly fits the criteria, you know? We can't make tiny digital cameras. And buying them premade would probably be so expensive we wouldn't make any money when we sold the finished gadget."

"They would be perfect!" Amber persisted. "And super popular!"

"But what about our goal? We're doing this to make money," Kammie reminded her. "I don't think we could make money with that, do you?"

"You're so logical," Amber pouted. "And probably right."

"And the lollipop!" said Trixie firmly.

They paused.

"A lollipop?" Wren asked. "Seriously?"

"A SPY lollipop!" Trixie nodded. "You know, like a spy would use."

"I don't have any idea what you're talking about," Wren rolled her eyes.

Winking at Wren, Kammie wrote it down and said, "That'll be *your* project, Trixie."

Trixie nodded back with a serious expression on her face.

Kammie moved to a fresh spot on the paper and wrote PROJECT IDEAS. Underneath she copied down the projects they'd decided on. It read:

- Code maker/decoder
- Periscope
- Spy Lollipop
- Magnifying glass
- Utility belt/gadget holder

- Entry detection kit
- Combination safe
- Invisible ink secret message

By the time they finished compiling their list, they could see the golden light of early evening spread over Wren's backyard through the big wall of glass. Trixie wandered back to the house for her bath.

"That took such a long time!" Wren complained. "Now we can't build any of these today. Just... promise me we'll actually make stuff next week. Look, that's even the next steps here. Create, then Test and Evaluate. Then we can 'Communicate the Results,' which for us means sell them and start making money."

Amber, Ivy, and even Kammie nodded at her.

Right now, though, they were worn out. All that thinking was surprisingly tiring. Ivy grabbed her soccer bag and led the way towards the back of Wren's house without a word, pulling out her phone to call her mom. The rest of them didn't have phones yet and had to ask Wren's parents to call for their rides.

Amber pulled on her cardigan and ran a finger around the empty popcorn bowl. As she sucked the salty butter off her finger, she said, "That was really helpful."

Wren had to agree. She had gone from feeling like a complete failure to feeling like she'd actually solved a problem or two.

The smell of dinner floated out as Ivy opened the sliding glass door. Grabbing her *Sarah and Simon* book, the popcorn bowl, and the empty lemonade jug, Wren dashed off towards her house.

"You were right, Kammie," Amber said. "Maybe brainstorming and making lists and plans wasn't the most glamorous stuff to do, but now we all know what we're doing. That's gonna make everything easier."

With a sideways grin, she put a hand on her friend's shoulder. "Now that we have a plan, what could possibly go wrong?"

What's
The
Problem

Imagine
(Brainstorm)

ENGINEERING
DESIGN
PROCESS

Evaluate

Create

Test

Share
The
Solution

9

AXEL ANDREWS & MILO JONES

*M*onday morning came and Wren was back at school heading to science class. The election-amplified chaos wasn't the only thing making her uneasy this morning. Friday's impulsive non-Newtonian flinging had put her on the wrong foot with her favorite teacher, Mr. Malcolm. But after her success over the weekend, she was determined to get back on his good side.

Once she figured out how.

She plodded past the candidates, the noise, and the bright colors on her way to the fourth floor, unceremoniously pushing the fliers away. The election showed no signs of settling down. In fact, she'd passed Benjamin Spencer in the hallway that morning and had to physically shove her way through the swarms of kids in his entourage.

But this morning she was lost in thoughts of how she could solve the problem with Mr. Malcolm. She squeezed through the kids around her locker without seeing them. Grabbing her science books, she started down the hall.

And almost got a mouthful of the perky blond ponytail of Axel Andrews. Wren weaved to the side, trying to slip by unnoticed. Chatting with Axel was the last thing she wanted to do, even before she'd tossed glop in her hair.

"Oh, hi Wren. How ARE you this morning?" Axel surprised her with a bubbly greeting, putting Wren immediately on her guard. "Who are you voting for?"

So that was her angle. Wren had forgotten Axel was running for a seat. She must really want votes if she was willing to talk to Wren.

"Move," Wren mumbled uncomfortably, forcing herself to add, "please." She put her head down, trying to step around the other girl's enormous, smiling white teeth.

Axel's grin took on a harder edge as she blocked Wren's way, "You really should vote for me, you know. Not only do you sort of OWE ME, I'm the best candidate by far. I honestly think I'd be really good at counseling!"

Wren rolled her eyes. "You certainly know how to order people around."

Judging from Axel's face, the words she meant to keep in her head had accidentally slipped out of her mouth. Whoops! She took a deep breath and tried again.

"Excuse me. Please. I'm late for class." Wren dodged around Axel and headed down the hall. While ducking another flier, she glanced back.

A boy Wren hadn't noticed stood next to Axel holding a tray of lemon cupcakes. Axel leaned towards the boy and said loud enough for Wren to hear, "I KNOW, she's totally rude, right? I still try to be nice to her, because if I don't, who will?" Then her voice faded into the din.

Wren hadn't been offered a cupcake.

A wave of insecurity hit her. *It's only a cupcake*, she told herself, *you're not even hungry*. And Axel had plenty of reason not to waste one on her. But she couldn't shake the feeling she'd failed some test she didn't even know she was taking. Her heart sank as she continued down the hall.

Sliding into her seat next to Amber, Wren dropped her head onto her crossed arms. Amber didn't seem to notice. She started babbling brightly the moment Wren appeared.

"I wonder what we're doing today? Mr. Malcolm isn't here yet. Hopefully he's not still mad. You know, Bobby says we're doing another experiment but I think it'll probably be a lecture. Maybe we'll be ma..." Amber's melodic voice suddenly stopped as the door to the classroom opened again.

Milo Jones walked in with his wavy black hair and warm brown skin, greeting Wren with a smile and a wave as he passed by. He didn't seem to see Amber but her eyes lingered on him.

With difficulty, Amber turned back to their table. She seemed to have lost her train of thought.

Wren quickly pulled the latest *Lovelace Gazette* from

her backpack and made a show of flipping it open, strategically placing it in Amber's field of vision, blocking the table where Milo sat down next to Bobby.

Wren selected the first distraction she saw. "Oh look! *Wicked* is coming to town! Think we can get tickets? You love that play!" The blurb was from "Axel About Town," the society and events column. Axel might be annoying, but she did have the pulse of San Francisco's best happenings for tweens and teens.

Before Amber could reply, Mr. Malcolm entered. He started talking before he'd even gotten to the front of the class. "Mrs. Yang's substitute and I are mixing up the classes today. We're going to compare and contrast what you've learned about states of matter. Mrs. Yang has been doing different projects than we have, with the obvious exception of last Friday."

Wren cringed and looked away. Then, she reminded herself she was going to get back on his good side, and refocused her attention on him with a smile.

"You'll have assigned groups," he continued. "I've put your groups up on the board. If your name is on this list here, you'll be trading places with Mrs. Yang's students. If this is you, grab your stuff..."

Wren's eyes widened as she read the names.

"Mr. Malcolm!" she called out impulsively.

He turned to her with a disapproving look, obviously

still mad. "Hands, please, if you have any questions," he said.

Wren clamped her mouth shut in a pucker and shot her hand into the air, waving it vigorously.

After a pause, Mr Malcolm asked in a voice dripping with impatience, "Yes, Wren?"

"I think," she started awkwardly. Then pointed to the board "I mean, is it... is that really my group?"

"Yes," Mr. Malcolm barely glanced at the board. He knew just what Wren was talking about. "And I expect you to be on your best behavior. Maybe even take the opportunity to apologize to Ms. Andrews."

Wren sighed, once again reading the names Wren, Milo, Amber, and Axel.

Axel entered with the group from Mrs. Yang's class and groaned after reading the names on the board, apparently as displeased as Wren with the seating arrangement. Amber, however, was uncharacteristically quiet. Milo joined them, flopping his notebook on the table and brushing long, wavy hair out of his soulful black eyes in a way that even Wren had to admit was pretty handsome.

Axel addressed Milo behind her hand conspiratorially jerking her head at Wren and Amber, "I guess we're stuck in this group, huh?"

"Yes, apparently," Milo looked straight at her instead. "Stuck like a duck in the muck."

Axel blinked in surprise.

Milo looked down at his notebook and began looking through his notes.

Wren snorted at Axel's confused expression, earning a carefully concealed smirk from Milo.

Amber, oblivious, could only stare at Milo and whisper, "Hi."

Milo nodded in her direction with a grunt.

Then Wren saw Mr. Malcolm watching her. She took a deep breath and turned to Axel. "Hey, so, I'm sorry I shot your head on Friday. It was an accident."

Milo, still pretending to look through his notes, broke into a mischievous smile.

Axel sniffed. "You know, I had to go through the whole day with a stain on my shirt. And my hair was ruined."

Milo looked bored but Wren tried hard not to roll her eyes. Instead she put on as honest a look of sympathy as she could manage.

"Actually," Amber found her voice again, "Oobleck is just water and cornstarch. Cornstarch is actually really good for oily hair, keeping it silky but still moisturized while soaking up the excess sebum from an oily scalp. It's also great for exfoliating to get rid of dead skin and to stimulate blood circulation."

"Really?" Axel perked up and turned her attention to Amber. "Like you just rub it on?"

"Sort of," Amber said, her eyes lighting up.

Wren zoned out as the girls began talking about the beauty properties of kitchen substances. She had made her apology and Axel was clearly not snarking at her anymore. That was as close to forgiveness as Wren expected.

The two other girls chattered on about the properties of apple cider vinegar for skin versus hair, and the healthy fats in avocados. Milo, sighing on the other side of them, caught Wren's eyes. They shared an eye roll.

Wren tapped her notebook in an unspoken question. He nodded with gratitude and picked up his pencil. As Axel and Amber moved on to the beauty wonders of used coffee grounds, Milo and Wren talked across them, discussing the actual science assignment.

Mr. Malcolm looked at their table and nodded approvingly. Wren caught it out of the corner of her eye and relief flooded through her. One problem solved, anyway!

10

SPIES

*W*ren sat in the Greenhouse on Saturday, waiting for the other Renegades.

It seemed like they were always planning their meetings around Ivy's games. Soccer, volleyball, basketball, whateverball. She understood how exercise could make you feel good and focus better, but Ivy was always at some practice or game. Wren just wanted to get started. She was doing her best to wait, but if they didn't show up soon, Wren would just have to start making stuff without them.

Wren was looking over last weekend's list of possible projects hanging on the easel when she finally heard the side gate creak and slam, followed by Ivy and Kammie's voices from the backyard. Kammie's mom had driven them both to Wren's house after Ivy's game. Kammie had never played any team sport, but she sometimes liked to watch Ivy's games. Besides, Ivy's mom had a conference call. If

Kammie hadn't given her a ride, Ivy wouldn't have been able to come.

"Yes, fruit," Ivy was saying. "Like an apple on a low tree branch."

"Weird," shrugged Kammie, opening the Greenhouse door for Ivy.

Wren looked up from the list. "What's weird?"

"Low hanging fruit," replied Ivy, smelling sweaty and setting down her soccer duffel.

"What fruit? Like a kiwi? Any fruit with fur is pretty weird," Wren agreed.

"No," Ivy laughed. "It's a business thing my mom was telling me about." Her mom was always saying adulty business things that Ivy repeated like she understood what they meant.

"Business fruit?"

"Low hanging fruit means an easier part of a bigger project," Ivy elaborated. "Mom was saying doing something easy when you're stuck can get you going again. Get you over the hump."

"With baby steps?" Wren asked. "That makes sense. Like putting your clothes in the hamper when you need to clean your room! It's easy because everything all goes in one place and then the room always looks a lot cleaner, so I feel like I've gotten stuff done. And it makes it easier to keep cleaning."

"Yeah," Ivy nodded while pulling up a stool. "I think so. Like putting all your dirty clothes in a laundry basket."

"Oh. Yeah. Well most of them are dirty, I guess. But it's easiest to just put them ALL in the hamper."

Kammie and Ivy looked at each other, but then they saw Amber heading towards them wearing a fancy blue dress with a full, swinging skirt. Her mouth moved silently as she lip-synced the words to some song in her head. Her small, graceful body slid across the yard in a complicated series of dance moves; spinning, stopping, and throwing out her arms. She suddenly stopped with her eyes closed, head back, and lifted her arms to the sky, belting out "...THE WAY IT GOES!" at the top of her lungs. Opening her eyes and waving through the windows, she lifted the corners of her skirt in a curtsy.

"It's such a lovely day!" she sang as she pushed through the wisteria and stuck her head in the door. "Shame to spend it inside!"

Tossing her soft white cardigan onto a stool, Amber spun to face them.

"I may have some trouble getting into it today. The trees are singing, the sun is shining, and the sky is blue!" She flopped onto the stool, propping her elbows on the table trying to look serious. "But! Here I am, ready to rock! Let's do this. What are we doing?"

Wren patted the list. "One of these. We're going to hang our fruit low and pick some easy ones to get started," she nodded to Ivy knowingly. She could speak adult business too.

"Um, yeah," Ivy lifted an eyebrow. "Don't think that's how you use that expression. Anyway, let's start with easy ones to get the ball rolling. I was thinking about the periscope last night. I'll tackle that one."

Kammie pointed to the decoder with a gleam in her eye. "I want to try this one. I love codes."

"I was thinking about codes too," said Wren. "Want to work together on it?"

"Yeah, absolut..."

"AAAAAAAA!!" Amber screamed.

Everyone's head snapped around.

She stared out the windows, pointing at a spot in the wisteria plant outside, and suddenly started laughing. "Sorry, scared me!"

Everyone looked where she was pointing. There, hidden in the overgrown leaves, two small round circles pressed against the glass. They were mysteriously attached to a pair of small hands and a fluff of light brown hair.

Wren squinted. The circles appeared to be the bottoms of a pair of binoculars.

As soon as everyone started looking at them, the binoculars and the hair fluff quickly faded back into the dense leaves.

A small voice called out, "You can't see me!"

The other girls laughed, but Wren twisted her mouth to the side.

"Of course we can, Trix. What are you doing?"

Trixie marched through the Greenhouse door defiantly, stuffing tiger-striped plastic binoculars into a canvas messenger bag slung across her shoulder. The army green bag looked enormous bouncing against her little body even though the shoulder strap was tied in a clumsy knot to make it shorter.

"I'm spying," she said with her chin up proudly. "And you can't catch me, I'm the White Mouse."

"The who?" asked Amber.

Trixie's eyes lit up as she pulled a large hardcover book out of the bag. She slid the book onto the potting table in front of Amber and pointed to a word on the title. S P Y.

"I saw this book in mom's office and snuck it for us, you guys. Look! I read this WHOLE WORD all by myself!" She gazed up at Amber, then turned to the rest of them. "It says SPY! I showed mom and she said I could bring it out here."

"Snuck it, huh?" Wren smirked as Trixie searched the pages for something. "With Mom's permission? You sneaky mouse."

Trixie found the page she was looking for and jabbed her finger on a black and white photo of a woman wearing an old-fashioned military uniform, smiling cunningly at the camera.

"That's Nancy Wake, the White Mouse. Mom read to me about her. She was super sneaky like me."

Wren put her finger on the page, then closed the book over her hand to look at the title. "*Spy Stories: 20 Incredible Women from World War Two*. Huh. Interesting."

She flipped back to the page on Nancy Wake and skimmed it quickly.

"You're right, Trix," Wren said. "Nancy Wake was a nurse from New Zealand who married a French guy. She was living in France when World War Two started. When the Nazis took over France, she started carrying around secret messages for the resistance."

"What's the Nazis?" Trixie asked.

"Bad guys," Amber replied simply, pulling Trixie onto her lap so they could look at the book together as Wren continued.

"No one could ever catch her, so they called her the White Mouse."

"Hmmm, and that's who you are today?" Amber ruffled Trixie's hair.

"Nancy Wake ran a whole network to get people in danger away from the Nazis. She had a five-million-franc price on her head. Franc?" Wren looked up.

"Old fashioned French money," Ivy explained.

"Wait," Trixie was confused. "Her head? What about the rest of her body?"

"It's just what people say," reassured Amber, giving her

a squeeze. "It means all of her."

Wren ignored them and continued to read. "The White Mouse soon led teams of up to seven THOU-SAND guerilla fighters. Wow!"

"That's a lot of gorillas!" Trixie said in amazement. "Was Bobo one of them? She must have needed a lot of bananas."

Ivy grinned at her. "Nah, this was real life, not *Sarah and Simon*. And it's not monkey gorillas..."

"Apes," corrected Amber.

"Ape gorillas. Guerilla fighters is spelled differently."

The book had a glossary and Wren, curious, looked up the word. "A guerilla fighter is someone who isn't in the army or navy or anything, but fights anyway. A human fighter. Apparently they fight differently than regular soldiers, with more sneaky strategy and surprise attacks."

"Oh," nodded Trixie. "That makes a lot more sense, I guess. Amber, come play spy with me!"

"Shoo!" Wren pointed to the door.

Trixie stuck her tongue out at her sister and grabbed Amber's hand, tugging her towards the door.

Amber laughed with a shrug, shooting an apologetic look at the other Renegades. "I'm not very inspired today anyway." Then she whispered loudly at them behind her other hand, "I'll keep her out of your hair while you work."

She let Trixie pull her out into the sunshine. "Let's go, little mouse. Let's be spies!"

11

THE PERISCOPE BUILD

*I*vy started with cardboard.

Last night, while reading *Sarah and Simon: The Seven Carat Conundrum*, Ivy had an epiphany about how to make a periscope. If she hadn't been thinking about periscopes already, the chapter might not have led to the epiphany, since it actually had nothing to do with periscopes at all.

In the chapter, the twin spies pursued Dr. Dinklemeyer's agent, Claude, deep into the desert to find a stolen seven carat diamond. The sun was just setting as they parked their car at a lonely outpost along the desert road. But there was no sign of the counteragent. All they found was an old mine shaft; really just a long, straight hole disappearing into darkness. Sarah pulled out her trusty grappling hook while Simon grabbed the large, silvery sunshades from the front and back windows of the car.

Strapping one sunshade to her back, Sarah climbed down into the dark mineshaft. Simon, up at the top, caught the light from the setting sun with his silver shade and angled it until the reflected light lit up the dark hole his sister was descending into.

When Sarah reached the bottom, she unstrapped her own shade and angled it to catch the light from Simon's. Then she moved it around like a flashlight, checking out the bottom of the mine. The missing diamond sparkled from its hiding place as the reflected sunlight struck it.

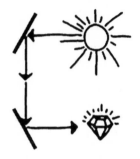

Ivy imagined her periscope working like that.

Instead of a mineshaft, she'd make a cardboard tube. Instead of the reflective sun shades, she could use aluminum foil. And instead of the sun bouncing around like a ball, reflections of what she was looking at would bounce around. Simple! She'd be done in a few minutes.

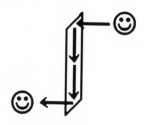

She grabbed some cardboard and plugged in the hot glue gun, setting it in the safe area they reserved as the glue station. Then she dug a roll of aluminum foil out of a bin and set it aside for later with a pair of scissors.

"Plugging in the gun!" she called out, so everyone would know to be careful.

Wren held up a hand in an "okay" sign but didn't look up from what she was doing.

Ivy was ready to assemble!

Folding the cardboard into a rectangular tube should have been easy, but at first it wouldn't budge. She pushed harder and it folded in a wiggly, angled crease. Ivy frowned. She wanted a nice straight edge. Unfolding it, she tried again in a new spot, this time bracing it better with her hands. The cardboard still wiggled down the middle of the new fold.

"Well, that's not going to work..." she mumbled, unfolding it again and smoothing it out. "Why is it doing that? Wait..."

She tried again, holding the cardboard against the edge of the table to keep the crease straight. Nope. And now her cardboard had a bunch of creases wandering across it like rivers across a map. Nothing professional or clean about those folds! Unusable.

Ivy growled, slapping it forcefully onto the card table. This was supposed to be the easy part! How hard could it be to fold a stupid piece of cardboard?

Behind her, working on their decoder, Kammie and Wren looked up.

"You okay, Ivy?" Wren asked, offering her a cheese stick. "Need some brain fuel?"

Ivy pondered the cardboard mess. "No thanks, I just have to try something new."

Wren shrugged and opened the cheese herself, sinking her teeth in absentmindedly as she turned back to her own project.

"Good luck!" Kammie offered a supportive thumbs-up, but Ivy was too focused on her project to notice.

Ivy sighed at the wrinkled cardboard, putting it down and looking out into Wren's yard through the windows. Amber and Trixie were playing Spy Tag. Amber snuck up to the very thin tree Trixie hid behind. It wasn't a very good hiding place. Even Ivy could see Trixie from inside the Greenhouse. Suddenly Trixie dashed away into the yard, laughing. Amber made a quick turn and caught her wrist just in time, changing her trajectory and spinning her around, Trixie's hair flew out in a circle behind her. Amber said something Ivy couldn't hear and Trixie nodded enthusiastically. Trixie put her hands over her eyes, fingers spread wide enough to see through, and counted. Amber looked around before running off towards the corner of the house and darting behind it.

Ivy took a deep breath, head a little clearer, and turned back to her cardboard. She held it up to examine it more closely. The cardboard was corrugated, meaning it was made of two thin sheets sandwiching a wavy line of some kind. All the scrap cardboard in the Greenhouse was from old, cut up shipping boxes. Ivy couldn't see exactly what was waving between the thin top and bottom sheets of her corrugated cardboard, though. Her folded piece was mangled and useless anyway, so she tore the top layer off. The middle layer was a third thin sheet just like the top and bottom, but glued in waves of long, even ridges. The ridges only went in one direction, like a bunch of straight rib bones. No wonder it was hard to fold against.

But it would probably be easy to fold ALONG.

She grabbed a new piece and made sure to fold it parallel to the ridges that time. It folded easily in a nice clean line.

Two more quick folds and she had a rectangle. She hot glued the edges together, making a long square cardboard tube. After slicing through all the corners at both ends to make flaps, Ivy folded one flap at an angle, and cut off the opposite flap. Then she hot glued the diagonal flap to the side ones.

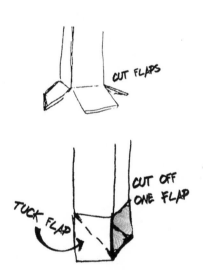

She stuck little aluminum foil squares on the inside of each angled flap. The thick hot glue made them a little lumpy, but she smoothed them out as best she could.

Ivy held up her periscope.

It was just what she had imagined!

Eagerly, she held the periscope to her eye, expecting to see a clear reflection of the other side. But all she saw was an amorphous blob of indistinct colors and light. She squinted, trying to see Wren and Kammie in the blurry foil. It worked great as a light bouncer, like Sarah and Simon's sunscreens, but as a periscope it was useless. The foil made a crappy mirror.

Ivy groaned. She tapped the periscope on the table, thinking. What she needed were actual tiny mirrors. Without them, the periscope was useless, a waste of time and effort.

She'd failed.

"I need a break!" she fumed, tossing the periscope onto the table angrily. She stomped past Wren, who was wrapping some paper strips around Kammie's wrist for some reason, and headed out the door to play a round of Spy Tag with Amber and Trixie.

Ten minutes of running around in the sun and her mind was fresh again. Fresh enough to remember the little square glass tiles they had in one of the bins, leftover from Amber's bathroom remodel.

She found two of the small tiles and hot glued them right on top of the aluminum foil. She carefully wiped the mirrored surfaces clean while the glue cooled.

Holding her breath, she lifted the periscope to her eye. On the mirrored tile, she saw Kammie and Wren reflected clearly, not even noticing her. She ducked under the table and poked the end of the periscope up over the top. She could still see them. But they couldn't see her!

Kammie lifted her head and looked around. Finally figuring out where Ivy was, she waved at the periscope.

"Did it work?" she asked.

"Yes! It's AWESOME! I'm so glad I didn't give up!"

PERISCOPE

MATERIALS:

- The *Periscope Template* on Cardstock (trace from back of the book or download and print online template)

- Glue or tape

- 2 mini mirrors (1 inch square)

Cut on SOLID lines

Fold on DOTTED lines

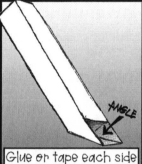

ANGLE

Glue or tape each side of the periscope into a rectangular shape, with one end angled and the other open.

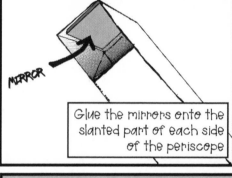

MIRROR

Glue the mirrors onto the slanted part of each side of the periscope

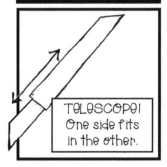

TELESCOPE! One side fits in the other.

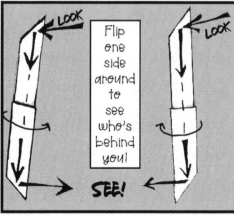

LOOK

LOOK

Flip one side around to see who's behind you!

SEE!

12

THE CIPHER CODE WHEEL

*A*t the exact same time Ivy began her periscope and Trixie led Amber outside to play, Kammie and Wren began working on the secret code tool.

Wren leafed through the *Spy Stories: 20 Incredible Women from World War Two* book while Kammie brought over some supplies. Behind them, Ivy had just picked up some cardboard. Wren stopped flipping through the pages when a photo of a woman caught her eye. The woman wore wire-framed glasses and her hair was curled away from her face, piled on top of her head. She wore an old-fashioned military jacket with stripes on the shoulder, a white collared shirt and a black tie. But it was her expression that captivated Wren. She looked like she had a thousand and twelve mysteries solved in her head but wasn't going to tell anyone the answers. She also looked a little sad.

"Who's that?" Kammie asked, looking over her shoulder.

"Her name was Joan Clarke, a cryptologist," Wren sounded out the long word while skimming the text under the image. "Says a cryptologist is someone who makes and breaks codes. Apparently, this Joan person worked with thousands of other women in England during World War Two to break enemy codes and decipher secret messages. She was so good at it that she worked in the highest group of codebreakers, figuring out the hardest codes invented and saving lots of lives. Wow! You have to be really good at puzzles to figure out codes like that. Sounds right up your alley, Kammie!"

Kammie pulled the book over and started reading. "She liked a lot of the same things I do. Language, logic, puzzles... She sounds really interesting! Oh! She was the longest serving member of her department, and she worked with some guy named Alan Turing to create a machine that helped break codes. Like, early programming or something maybe?" Kammie looked up at Wren, her brown eyes shining. "You know, if we were spies, you'd be a field agent, and I'd probably be a cryptologist like Joan Clarke. Do you think I could be a cryptologist even if I'm not a spy? I might be really bad at it, but it sounds like a fascinating job."

"Maybe," Wren shrugged. "I've never heard of cryptologistry before. Speaking of, you said you had an idea about the code gadget?"

"Yup," Kammie nodded, "At least a place to start. I'm sure you can improve it. You know how Simon has his secret decoder belt buckle? It works by shifting the alphabet one way or the other, and substituting the shifted letters for the real ones."

"Mmm hmm," Wren agreed. Agent Zero had given Simon the buckle on the twins' first adventure. "He unfolds it and it becomes two copies of the same alphabet, one row on top of the other. When he slides the top row over, it lines up with different letters on the bottom row. Then he writes a message using the top letters but the real message is in the bottom letters they match up with. Like if he shifted the top alphabet over three letters, an A would be a C, and a C would be an E. So CAT would be coded as EC... ummm... V."

Kammie nodded enthusiastically, "That's two letters, actually... My puzzle book calls that sort of code a 'substitution cipher.'"

"Right, two letters," Wren counted on her fingers. "What does cipher mean? Is it another word for code?"

"Pretty much. I think codes and ciphers are a little different, but I'm not sure exactly how. I've also seen a substitution cipher called a Caesar cipher, because an ancient Roman emperor named Caesar used them a lot. I thought we could make one of those. Like Simon's belt buckle but not metal. What do you think?"

"Sounds good to me," Wren shrugged.

They began by writing two copies of the alphabet on matching strips of paper, being careful to make all the letters the same size so they would match up. Then they slid the top strip over a few letters. Kammie, matching the real alphabet on the top with the shifted alphabet on the bottom, and wrote the code "WKLV LV IXQ" in her notebook. Wren used the strips to decode them to "THIS IS FUN."

Suddenly Ivy, who was working on the periscopes at the other table, growled and smacked down some mangled cardboard.

"You okay, Ivy?" Wren asked. "Need some brain fuel?"

Apparently, Ivy was not in the mood for snacks. "No thanks, I just have to try something new."

Wren took the cheese stick for herself and began nibbling on it absentmindedly.

"Good luck!" Kammie sent Ivy a thumbs-up and turned back to the cipher too. Behind them, Ivy continued working on her periscope.

"So one problem I see is here," Wren pointed out the beginning of the row. "These letters hang off the end and don't have anything to match to when you shift them, so you have to sort of wrap it around in your brain to figure it out. It would get harder the more letters you shift it over. Wait... wrap... that gives me an idea."

She picked up the pieces of paper and wrapped each of them around Kammie's wrist like a pair of alphabetic bangle bracelets. The letters lined up perfectly, so they matched when one bracelet was shifted.

"Oh that's so cool! And fashionable! Amber would love it," Kammie smiled.

Behind them, Ivy looked into a cardboard rectangle.

"But how do we keep them together?" Wren pondered. "If they don't stay together, they won't be a usable code tool."

Suddenly, Ivy said she needed a break and stomped out the door.

Kammie and Wren looked at each other.

"I know how that feels, when you want to give up," Kammie said, watching as Ivy chased after Trixie.

"Yeah, me too," Wren agreed. "I think everyone does! You just have to come back to it when you've cleared your head, right?"

Kammie nodded.

They tried a few different ideas to keep the alphabet-strip bracelets together but couldn't figure out how both bracelets could stay together but still shift around. Wren wadded up the bracelet strips and they started over.

Wren pulled out some stiff cardstock paper and cut two matching circles. They drew a bunch of lines across both circles to make twenty-six sections, one for each letter of the alphabet. It was hard to get the spacing just right because they had to be equal or the letters wouldn't line up when they got shifted. After some trial and error, they worked it out. Once both circles were divided into twenty-six wedges, Wren trimmed one of them down an inch. Since the wedge sections all started from the middle of the circles, they would still match up with each other no matter what size the circle was. Then Kammie used her neat handwriting to fill in each wedge on both circles with the letters of the alphabet. When they stacked them together and poked a brass fastener through the centers, they had two rotating alphabet disks that matched up perfectly.

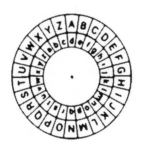

"Triumph!" yelled Wren, turning the top disk and writing "WULXPSK!" in code in her notebook. But when Wren gave the code disk to Kammie to translate her message, the top disk shifted.

"Wait," Kammie asked, "What was the code set to?"

Wren struggled to remember. "You know, I honestly don't know. I just turned it and wrote."

"Hmmm, that's a problem if we're making these to sell. People need to be able to pass messages without having to tell their partner the code, right?"

"Yeah," Wren mused. "Hey! What if, instead of matching two alphabets, we only used one alphabet. And covered it up? Then we could cut two little windows, right? One of the windows would show the real letter, and the other would show the coded letter."

"That could work!" Kammie nodded. "Let's try it."

By the time Ivy came back, the girls had recycled the top disk and replaced it with a full sized one with nothing written on it. Wren was just cutting two windows in the top disk as a flush-faced Ivy settled back into her seat with renewed purpose.

Once the two windows were cut, they realized there was no way to tell which was the code and which was the real alphabet. Kammie suggested marking the windows with two different colors so they could tell them apart.

Suddenly Kammie realized Ivy was gone again.

The door hadn't opened, so Ivy must still be in the Greenhouse. Wren nudged Kammie, pointing to a little cardboard square poking up from under Ivy's table. Kammie had completely missed it! She waved at the periscope happily.

"Did it work?" she asked Ivy.

"Yes! It's AWESOME! I'm so glad I didn't give up!" Ivy replied from under the table. "How about you guys?"

Wren showed her the code wheel.

"That's awesome, but can the codes be different? Like, can the windows be cut over different letters?" Ivy asked.

"Oh yes," Kammie pointed to the top disk. "You can cut them out anywhere, there are like twenty-five different code options, depending on where you cut the second window."

"So," Ivy considered the possibilities, "can we just cut one window when we sell them? That way whoever buys them can cut the second window themselves to have their own unique code that even we wouldn't know."

Right then, Amber came back in. Trixie held one of her fingers and trailed behind her.

"Check it out, guys! Two amazing finished spy tools!" said Wren.

"Is one of them the Spy Lollipop?" Trixie asked eagerly.

"Uh, no, dorkus. That's all you. We made a periscope and a cipher code wheel," said Wren, beaming. "If we can find a way to make a whole bunch of them without spending too much time and money on the construction, we could sell gazillions of them!"

"You know what we could do," Amber suggested after looking at the inventions. "I could put both these patterns into a computer, and print them on stiff cardstock. My Dad could help with his graphic design software. The cipher code wheel thing would be super easy and look a lot better with typed letters. In fact, if we're printing the periscope on a single sheet of cardstock, we could split the tube in two and make one side a little thinner than the other, and it would telescope!"

"Telescope?" asked Trixie. "Like what pirates use?"

"Ah, no, not that kind," Amber explained. "I mean like 'telescoping'... Get bigger by extending out, like a handle of a little umbrella. One half would sit inside the other, and when you slid it out, it would get longer. What do you think?"

"That's a fantastic idea! And making the periscopes out of cardstock would solve my folding issue. The corrugated cardboard was tough to work with," Ivy gushed. "Can you and your dad do that?"

"Put them in the computer and print a bunch?" Amber nodded. "I don't see why not. He's home today watching my brothers while Mom's at a meeting. We could use money from the club treasury to buy fancy colored card-stock paper, and I think I saw some little mirrors like these at the craft store. We could get a bulk pack! Then we could make a whole bunch of both of these gadgets before Monday!"

CIPHER CODE WHEEL

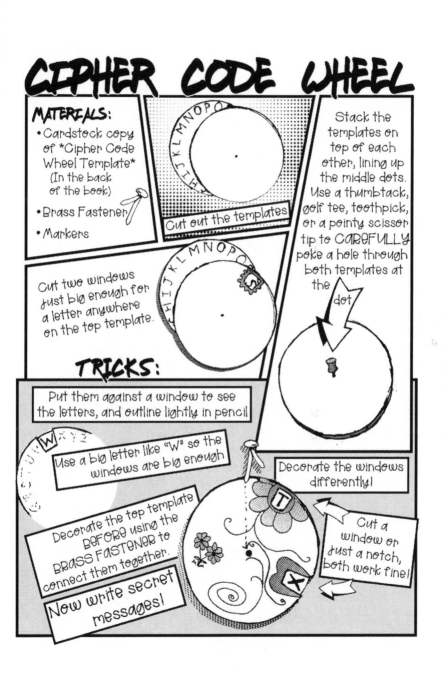

MATERIALS:
- Cardstock copy of *Cipher Code Wheel Template* (In the back of the book)
- Brass Fastener
- Markers

Cut out the templates

Stack the templates on top of each other, lining up the middle dots. Use a thumbtack, golf tee, toothpick, or a pointy scissor tip to CAREFULLY poke a hole through both templates at the dot

Cut two windows just big enough for a letter anywhere on the top template.

TRICKS:

Put them against a window to see the letters, and outline lightly in pencil

Use a big letter like "W" so the windows are big enough

Decorate the top template BEFORE using the BRASS FASTENER to connect them together.

Now write secret messages!

Decorate the windows differently!

Cut a window or just a notch, both work fine!

13

OPEN FOR BUSINESS

*A*s usual, Student Council candidates prowled the Monday morning crowd, greeting the voters with their insincere friendly smiles. But this Monday would be the start of something new. The Renegades were open for business! Almost. Amber had the inventions they'd tinkered up, and they planned to set up in the recess yard at lunch.

Moving smoothly through the wild halls, Amber barely glanced at the campaign posters stuck to the lockers. She accepted a cookie from a candidate she didn't know, humming to herself. She'd never remember his name come election time, but the cookie was delicious.

"Look out!"

The voice shattered Amber's reverie as she almost ran into Tiffanie, a sturdy girl with long brown hair.

"Sorry, Tiffanie," she sang, pirouetting out of Tiffanie's way, freckle spattered arms over her head like a ballerina.

Spinning around, Amber saw "VOTE FOR BOBBY" written in bright marker letters on the back of Tiffanie's shirt. Bobby was Tiffanie's boyfriend. Amber wondered what it would be like to kiss a boy. Like Milo.

She also wondered if she should vote for Bobby

Before Benjamin Spencer, Amber just wrote "Wonder Woman" on her ballot. But not anymore. Amber wanted to vote for a candidate that would help make a difference at the school. Would Bobby make a difference?

Arriving at her locker, Amber carefully placed a box full of periscopes and cipher code wheels on the shelf, praying it would fit. It did, with a bit of encouragement. The locker door clicked shut with a shove. Amber pushed on the door experimentally, reasonably sure it wasn't going to pop back open. Since the lockers at Lovelace didn't have locks, their contents sometimes exploded into the hall if they were overstuffed.

The first half of the day ticked away slowly. When the kids were finally released for lunch, she flew to her locker. Gathering the box with the spy gear, she sprinted to the recess yard.

Dark green wooden picnic tables lined the edges of the yard. They overlooked the ball court and squishy-floored play structure. Kids were swarming the tables as fast as they could reach them, but only a few were lucky enough to grab one. A pack of players with Pokémon cards shared the table furthest from the action with a few kids reading their *Sarah and Simon* books.

Three student council candidates pretended to share another table gracefully. The athletic boy, apparently named Tyrone, and an older girl with short pixie hair were eying the other candidate with jealousy.

Fabulous multicolored glitter shimmered "Vote for Afsheen" around a large color photo of her shaking hands with Benjamin Spencer himself. Afsheen clearly understood how to appeal to the voters!

Amber, the first Renegade on the recess yard, managed to snag half a table close, but not too close, to the play structure. She quickly draped an aqua-colored satin cloth over the rough green wood, claiming a large chunk of the coveted table. Next to the cloth, she taped a sign made from a folded sheet of cardstock. It said "Super Spy Gear for SALE. Made by the Renegade Girls Tinkering Club" in swirling letters, full of curly-ques.

Kammie arrived and set a small spiral memo pad and a small lockable metal box on the table near the sign.

"This is our notebook to record sales, and this is our money box!" she proclaimed. "I keep the key around my neck so our money will be safe," she winked at Amber and pulled on a thin pink ribbon necklace around her neck, showing off a small key strung on it.

Wren showed up at the same time as Kammie, setting

out a few brightly colored periscopes, a handful of cipher code wheels with only one window cut out, an empty notebook for example codework, and two samples of each spy gadget clearly marked SAMPLE on the side. They wanted to make sure they kept some for demonstration purposes.

By the time Ivy arrived, the girls had everything ready to go.

"Hang on," Ivy evaluated the set up. "We better spread out. So we have enough room to talk to all our customers at the same time. I'll take this corner."

Amber, taking the front corner opposite Ivy, looked at their display with satisfaction. She rubbed her hands together and turned, looking eagerly around at the sea of kids racing across the playground.

"And now," Ivy said, hands on her hips, "We wait!"

And wait they did.

A few kids glanced over curiously while running past, but otherwise no one seemed to notice them.

Ivy caught the eye of a group of giggling third graders, preparing to launch into her sales pitch. But they walked past, nodding at her politely.

They waited some more.

Except for a few random kids who glanced over at their table, and one sixth grader who pointed them out to her friends, no one even seemed interested in the Renegades or their gadgets.

Eventually, recess crawled to an end and everyone started to head back inside. They hadn't sold a single item.

Slowly, Wren packed everything back into the box. Amber shook out the shimmering cloth and slipped it into her back pocket morosely. Kammie picked up the empty money box and shoved it forcefully into her backpack. She sat down with a heavy flop on the top of the picnic table, sniffling back tears, and grabbed menacingly at the cardstock sign, mumbling something about failure.

Ivy swiped the sign before Kammie could crumple it and put her hands on her hips with a determined expression. "Cut it out, guys, it's just our first day! Don't get discouraged. We'll do better tomorrow."

But the next day wasn't any better. The Renegades spent their whole recess rearranging periscopes into various color patterns and twisting cipher code wheel disks. Wren finished some math homework, so at least it wasn't a complete waste of time.

"I don't get it," grumped Amber. "It's good stuff and we worked so hard on it. Why isn't anyone interested in buying our spy gear?"

Kammie decided not to remind them that she had warned them starting a business might not be as easy as it seemed. She was tracing her finger around some initials carved into the picnic table when a shadow covered her hand. She looked up into Axel's light blue eyes. Axel had two kids with her, a girl with braces, and a boy carrying a tray of cupcakes.

"What's all this?" she asked curiously.

Kammie just stared at her, unable to make herself answer.

Axel frowned at Kammie's silence and turned to Amber. "What is all this?"

Amber took a deep breath, "Oh, hi Axel. We're selling spy gear to raise money for our club. Want to take a look? These periscopes here let you see around a corner, and these help you code and decode secret messages."

Axel looked interested when she saw the periscopes. When she looked around and seemed to suddenly notice no one else was at the girls' makeshift shop, her reaction changed. Her mouth twisted to the side.

"Ummmm. Sorry guys, I have more important things to do." And off she went with a flip of her ponytail, laughing with her friends while throwing a quick look back.

No one else stopped by.

Near the end of recess on the third day, Amber was ready to give up.

"I'm bored!" she moaned, laying across the picnic table next to the satin cloth, her head hanging off the end. "This was a stupid idea, and we wasted so much time on it. We should have made lemon bars. If those didn't sell, at least we'd have something to eat!"

She watched the kids shooting hoops and running,

climbing, and spinning on the playground equipment. From her vantage point, with her head dangling off the table and her long hair blowing gently in the breeze, the other kids looked upside down. Watching the climbing structure upside down was particularly intriguing. As she watched, a lone figure walked upside down towards her. Amber suddenly realized the figure was coming to their table and showed no signs of going around. She sat up so quickly that her vision blurred. She swayed dangerously on the edge of the table.

A strong, steadying hand touched her shoulder.

"Look out, there!" said the voice attached to the hand.

Amber realized none of the other girls had said anything. They seemed frozen.

Confused, she turned to the voice and said, "Thanks!"

"No problem," Benjamin Spencer replied.

14

THE TASTE OF SUCCESS

"Would you, ummmm, like to see some spy gear?" Amber hopped off the table, giddy with awe. She was talking to Benjamin Spencer! "We are selling some fine wares." And she was being a dork.

"'Fine wares'?" Wren whispered to Kammie behind the back of her hand. "Seriously?"

Kammie was staring at the ground quietly, as usual. She barely managed a shrug.

"Yeah, actually," Benjamin replied. "I've been seeing you guys over here. I'm curious. What do you have?"

Other kids were glancing over at their table. No one else came over, but more than one group of kids suddenly seemed to be considering it. A red-haired boy in the middle of a large group pointed.

Trying to stay focused and land their first sale despite

all the distractions, Amber explained about raising money for their club. She told him how the periscope worked, demonstrating with the example.

She and Benjamin crouched down and used it to look over the edge of the table together, and she showed him how he could turn the extending part around so he could look behind himself. That was a little trick they'd discovered while sitting around.

"Cool!" he replied with genuine enthusiasm. "What's this other thing?"

So Amber showed off the cipher code wheel, using the blank notebook to encode and decode a message with one of the samples. She showed how the code was different with the other sample's window cut over a different second letter.

"You can cut the second window anywhere you want, and it'll be a unique code! Then just give a matching decoder to a friend and you can communicate with secret messages all day long!"

Benjamin was truly impressed. "How much are they?" He asked while pulling out his wallet.

Kammie had unlocked the money box during the sales pitch and set it near Amber, then silently moved a few steps behind Wren, watching over her friend's shoulder, owl-like. From behind Wren, she pointed to a price list she had made yesterday. Each letter in "Price List" was filled in with a different color, like a rainbow. Benjamin grinned at it.

"Do you have change?" he pulled out a large bill.

Amber, who'd been excitedly gathering up his items, paused and looked up at him sheepishly.

"Change?" she asked.

"Yeah, you know, so I can buy some."

Amber's face fell. "We, ummmm. We didn't think of that."

Her thin shoulders drooped at the loss of their first sale. Ivy looked away with a frown. Amber shrugged at Benjamin with a helpless smile as she turned the empty money box upside down.

"Thanks anyway," she said in a sad voice.

But Benjamin just laughed. "Hey now, don't get so discouraged! New ideas are never easy. These are actually really neat. Why aren't you advertising? I bet if the other kids knew they were here, they'd buy a ton!"

The girls looked at each other. Advertising? Change? Kammie was right. Opening a business was a lot more complicated than it seemed.

Benjamin laughed again.

"Tell you what," he said, glancing again at the price sheet. "I'll just take two periscopes and five cipher wheels. Then you don't need to give me any change. I can give the decoders to the other members of the student council. It'll help get the word out, and we can use them to pass secret student council notes around when we need to communicate about private business."

Amber was flabbergasted. "Wha..." she blushed. "That... That would be amazing! Really?"

Ivy stepped in and gathered up the requested gear, handing it to Benjamin herself. Amber took his money, dropping it ceremoniously into the no-longer-empty money box.

"Well, enjoy them!" Amber told Benjamin, walking a few steps away with him as he left. "Thanks for your patronage!"

He winked at her. "You bet, thank YOU."

After a few steps he turned back around, looked Amber in the eye and said with a grin, "You know, you're pretty smart. Have you ever considered running for student council?"

Amber laughed. Then, lost in thought, she wandered off to class, leaving the other Renegades, her satin cloth, the money, and the rest of the gadgets behind.

15

BETTER AND BETTER

*A*fter Benjamin's visit, business picked up.

Kammie's mom swapped out the big bill with lots of smaller bills for change. But advertising no longer seemed necessary. Word spread quickly and by the end of the week, the girls had sold out of all their periscopes, and all but two of their cipher wheels.

By Friday, the girls' moods had picked up too. Wren hadn't even been sent to the principal's office all week. As they sold their last periscope, another unexpected customer stopped by: Gail Mendez. Gail, the eighth grade editor of the school newspaper, rarely did her own reporting these days. She was too busy editing and running the paper. But Gail Mendez herself stood at their picnic table, holding a spiral notebook and a smartphone.

"Hey ladies," she already had their full attention. "Can

you spare a few minutes for the school paper? I'd like to interview you."

None of them had ever been interviewed before. Not sure what to expect, they agreed to meet Gail during afternoon break. That gave them about an hour to get really nervous.

As she headed back to her locker, Wren noticed Axel watching them out of the corner of her eye.

"So," asked Gail near the end of the interview. She flipped over to a new page in her top-bound spiral notebook. "What's next for the Renegade Girls Tinkering Club and your spy gear? What fun products do you have coming up?"

The girls looked at each other, except Kammie, who had spent pretty much the whole interview silently looking at the hands in her lap. The art room was a quiet haven. The girls weren't sure how Gail had secured it for the interview, but they had shown up as requested during afternoon break.

"Well," Ivy began, continuing to pull the bristles out of the paintbrush she'd been fiddling with for the last ten minutes. "We haven't really... I mean... there are a bunch of ideas on our drawing board."

"Oh?" Gail asked with interest, jotting down some

notes. She brought her pencil to her mouth and chewed on the eraser. "Proverbial or literal drawing board?" she joked.

"Literal," Kammie answered, surprising everyone. "We use an easel."

It was the first time she'd spoken during the interview, and Gail stopped writing to listen with renewed interest.

"Really? That's actually pretty cool," Gail indicated her paper notebook and pencil. "I like writing stuff on paper, too, before typing it up for the newspaper or my home-work. I just remember things better when I do. So what do you have on your easel right now?"

"Well, I don't..." Amber said haltingly.

"I mean we haven't..." Ivy replied at the same time.

But they both stopped as Wren leaned forward over the table with a mischievous half smile. Gail's attention shifted to her as Wren waited with a short, theatrical pause.

"You'll just have to wait and see," she answered mysteri-ously, wiggling her eyebrows.

"I look forward to it," Gail chuckled. "Alright, thanks so much for your time, ladies! Good luck to you. Make sure you check out Thursday's edition of the paper. My article should be up by then."

She picked up her smart phone and hit stop on the recorder.

As Gail gathered her things and turned to leave, Amber hesitantly reached out a hand.

"Oh, wait," she said timidly. "One more thing."

"Yes?" Gail asked, looking back over her shoulder.

"Ummm, could you mention that I'm running for student council?"

Ivy, Wren, and Kammie all stopped packing up their own notebooks and looked at Amber.

"You are?" Gail asked.

"Yes," Amber replied with growing confidence. "It's sort of a new thing, so I'm a bit behind."

Gail grinned. "Hijacking a news story to get your name out, eh? Cunning! I like it! I'll see what I can do."

Gail winked at Amber as she strode out the door.

There was a moment of silence as the color rose in Amber's cheeks. She lifted her jacket from the back of her chair and put her arm through a sleeve. None of the other girls moved except to watch her.

"Well," Amber ducked her head as she pulled the other sleeve over her arm and said, a little too loud and a little too fast, "I gotta go. I need to get to the front office to put my name on the candidate list before break is over. See ya!"

She opened the door of the art room wide and kicked the door stopper into place. The other girls watched her head off down the empty hallway.

"Wait, what?!?" called Wren as Amber walked away quickly. "Why didn't you tell us?"

Wren turned to Ivy and Kammie. "She didn't tell us,

right? I mean, I didn't forget something that big? Did she tell you guys?"

Kammie shook her head. "Not me."

Ivy brushed the pile of dislocated paintbrush bristles off the table into her hand and dropped them into the garbage can. Dusting off her hands she added "Nope. Me neither."

16

THE NEXT STEPS

"We're going to need to pick up some new supplies," Ivy declared. "If we're going to make new stuff!"

After the interview, the Renegades assembled for a quick emergency meeting after school. No one acknowledged Amber's bombshell candidacy, including Amber. Wren was still confused by the idea that her best friend could be interested in taking part in the chaos that was driving her crazy. Instead, they concentrated on growing their business. Gail's questions had gotten them thinking about new inventions to add to their inventory. They reviewed the list Kammie had copied into her notebook from the easel, and decided to make gadget holder utility belts and a secret message kit Ivy had an idea for.

Kammie copied some numbers into a hastily made grid

on scrap paper stapled inside her notebook. She turned to Wren, "Hey, can you total these numbers for me?"

As Wren did the math, Kammie explained. "So, according to these sales numbers, and projecting sales for new merchandise based on the previous sales, AND factoring in continuing sales for our current gadgets... well, just look at the numbers!"

Everyone crowded around Wren as the numbers she was adding up kept going higher. Wren added faster the further she went. Kammie started to giggle behind her hand and Ivy patted her on the back

"So we'll have enough to fix the microscope?" Amber asked.

The others paused.

"Well, sure," said Kammie, "It seems entirely likely."

"But we could do so much more!" Ivy looked at the final numbers with wide eyes. "If we continue this kind of business, we could fund our treasury for a year! By the end of the school year we could get all sorts of new stuff for the Greenhouse!"

Amber squinched up her face, "Yeah, but then what? When would we actually get to USE the microscope. And all those new slides?"

"We can use it any time," Wren shrugged. "And get even more slides."

"But we haven't even looked at the ones we already

bought," Amber objected. "I thought we were just doing this to fix the microscope eyepieces."

"Aren't you having fun building the spy gadgets, though?" Wren gazed at her dreamily. "I'm really enjoying inventing them."

Amber shrugged. "Sure, but I'm not sure that's all I want to do all year."

"I hear you," Ivy said, putting on her serious business face. "But we don't have enough money yet anyway. Let's keep going and see how it goes."

Kammie nodded. "You know, once the article runs in the paper, I bet the other kids will be really impressed. My projections might even be low."

"We'll have to use some of what we've earned to buy the supplies for the secret message kits," Ivy told them thoughtfully. "But I think they'll sell really well, so we'll make it all back and more."

Wren wiggled with excitement. "I'll ask my mom if she can take us to the electronics store in the mall tonight!"

WREN CRASHED in through the front door of her house, kicking aside a stack of Amazon boxes that were in the middle of the hallway.

"Mom! Moooom! Mom! Mom!!" she yelled as she

tossed her backpack on top of a pile of laundry on the couch. "Hey Mom!"

"What, honey?" her mom emerged from her bedroom, carrying an armful of long metal rods and trying to adjust the purse strap over her shoulder. She was wearing huge, stiff gloves that made the job difficult. "What do you need?"

"Where are you going?" Wren asked, temporarily side tracked, taking in the weird outfit in front of her.

"I'm headed to my welding class, honey. It's Friday," Her mom shifted the metal to one arm and adjusted a pair of heavy goggles that were on her head. "Do you know where my car keys are?"

She started to dig awkwardly in her purse, still wearing the heavy gloves, and dropped a few metal bars to the floor with a deafening CLANG! "Shoot! I know I put them in here somewhere!"

Wren reached over to the hall table and grabbed a thick ring of keys that were attached to a purple carabiner. She hooked the purple clip to the belt loop of her mom's jeans and stooped to pick up the metal rods.

Her mom gave her a smile of thanks and took the rods from her, struggling to balance them in her arms while she opened the front door. "Your dad will be home soon. His bus is running late again."

She paused on her way out and turned to look at Wren, "You'll be okay alone for a few minutes, won't you? Trixie is over at Rishi's."

"Sure, of course I'll be okay, Mom. I'm eleven for heaven's sake!" Wren replied, reaching out to hold the door open. "But I thought you were doing glass blowing."

"That was last month, sweetie," her mom said, angling the long rods out the door carefully and heading down to the garage.

Wren suddenly remembered what she was going to ask. "Hey Mom! I need to go to the electronics store with Ivy! I was hoping you could drive us."

"Not today, honey. Maybe tomorrow. Wait, shoot, I have that meeting. Maybe in the afternoon? Remind me and we'll figure something out. Okay? Bye! Be good. Don't forget the rules! See you soon," Her mom gave her a tiny wave while fumbling for the garage opener on her keychain.

Wren sighed. As she pulled the door shut, she heard the garage door rumble open. Then she heard her mom's Prius pull out into the traffic on their street and the angry honking of the other drivers, so she reached for the spare controller and closed the garage door her mom had left open.

The handset for their house phone lay on the couch on top of some junk mail. Wren scooped it up and dialed Ivy's cell phone number, flopping on the couch without bothering to move the mail. It wasn't fair that Ivy was the only Renegade with a cell phone. She was jealous. But then, Ivy was older than the rest of them. Over a year older than Wren. A lot could change in a year. Maybe this was the year Wren would finally get her own phone too. Right now she was just thankful her family still had a house line.

Ivy picked up. "Hey, Wren. Did you ask? Can your mom take us?"

"Nope, she just hurricaned out the door on some wild adventure," Wren shook her head. "She's headed to Oakland. Between her class and rush hour traffic she might not even be home for dinner. And Dad's running late because of stupid MUNI. I tell you, our public transportation system needs a SERIOUS overhaul. His bus is always running late."

"Darn it! I'll have to ask my mom. I was hoping I wouldn't have to," Ivy groaned. Her mom, an electrical engineer, was some kind of boss at her company. She didn't have a dad. Ivy's mom worked long hours to get everything done every day, and went out of her way to take care of everything Ivy needed. But a day only had so many hours and her mom was only one person. One person with a demanding job and lots of important projects to oversee.

"Sorry," Wren felt a pang of guilt. "I'm a victim of circumstance."

"Let me check," Ivy said with resignation. Wren heard her lower the phone and call to her mother in Korean. "Umma! Hey Umma!"

Wren zoned out as Ivy negotiated with her mom. Eventually, Ivy picked up her phone again.

"No luck today. We'll have to wait for Saturday during club time, after my soccer game."

Wren groaned. Solving problems was a lot harder when you had to rely on other people!

17

BESPOKE UTILITY BELTS

*B*eautiful fabric pieces covered the potting table in the Greenhouse.

Satin, cotton, stretchy t-shirt fabric, and even some velvet screamed "touch me!" as Wren ran her hands through the delicious colors. Reds, pinks, blues, silvers, you name it. Wren's mom, who was only happy when she was busy making stuff, had donated all her scrap fabric to the club. She loved to sew costumes and make fancy historical clothing reproductions when she wasn't welding, or woodworking, or doing some other random thing.

Wren loved the fabric. The way it shimmered, the riot of bold colors, and especially the feel of smooth, cool fabric slipping along her hands and forearms.

The fabric was for the gadget holders, or as Amber had decided to call them, fashion utility belts. By "utility belt," she meant something a spy wears to carry equipment, usually with special compartments for different tools. It didn't really have to be a belt. And Amber was excited about inventing them.

When Amber opened the door to the Greenhouse, Wren was tying a strip of dark green velvet over her eyes. She'd already cut two eye holes in it.

"I'm incognito, Amber. Going incogNEETo!" she said.

Amber had a fancy paper shopping bag with her. It said "BESPOKE" in almost indecipherably swirly lettering. Rolled up poster board in various colors stuck out of the top. She set it down on the potting table and reached for a handful of the soft, silky fabric.

"I had a great idea about the utility belts," Amber said, cushioning her face on the fabric. "What if we make bespoke spyware?"

Wren lowered the mask and raised an eyebrow. "Buttspank spyware?"

Amber stared at her icily, clearly not amused, while Wren snorted with laughter.

"Sorry," Wren coughed, settling down. "Bespoke, right? What the heck does 'bespoke' mean?"

"It's SUCH a great word!" Amber gestured towards her bag with enthusiasm. "I learned it the other day at this INCREDIBLE new store in my neighborhood called Bespoke! It means one-of-a-kind. Bespoke has these AMAZING clothes! Mom got me a top and two hair-bands, all handmade. You wouldn't believe it. The most gorgeous colors and fabrics and styles! And everything is unique. So, of course, it's all pretty expensive. If we make our utility belts unique, we can charge more for them!"

"You know, that's not a bad idea, really," Wren agreed. "I mean, it would be tough to mass produce them the way we do the periscopes and ciphers anyway, right? Can't put fabric in a printer! Trust me on that, by the way, don't even try it. So why not make them totally unique?"

Amber nodded, "Right? I bet they'd sell really well. I bet kids would be willing to spend more money on our stuff now that we have a good reputation. Reputation is everything in fashion..."

By the time Kammie's mom dropped her and Ivy off at Wren's, Amber had already set up the sewing machine and plugged it in. She perched on the stool, with the pile of fabric next to her, sketching in her notebook.

The bag of rolled poster board sat on the floor where she had moved it after pulling out the old sewing machine. Amber had intended to start creating some campaign posters, but had gotten sidetracked by what she was supposed to be doing in the first place, creating the merchandise.

"So what are you planning on doing with all this fabric?" Ivy asked.

After Amber explained her plan, Ivy nodded with approval, but Kammie looked concerned.

"How many do we have to make? Will they really be

unique?" Kammie's voice got a little higher with each question. "How different can one utility belt be from another? Do we have to brainstorm a whole new list just for them? What if we need fabric we don't have?"

Amber rested a hand on her friend's shoulder. "It's okay. Look, I've already sketched a few ideas. This one stores a few tools around your waist where you can get them quickly, like a classic utility belt. It'll be good for when you need your hands free, like when you're scaling the outside of a skyscraper using suction cups. This one here is a fancy sash, for when you're undercover. See how the pockets to hold stuff are hidden on the inside? Oh! And this one is my favorite. It looks like a regular purse, but when you flip open this secret side here, it's a bunch of pockets and straps with velcro to keep your stuff secret! Plus, you still have your lip gloss and money!"

She went on to show off other designs from her notebook.

Kammie was impressed but not convinced. "I'm not very good at sewing. I don't know how to do things this complicated."

"YET!" Amber added. "You just need more practice. But don't worry. I'm totally on this."

She shook out some light blue cotton to see how big the piece was. A spool of red thread dropped out and rolled across the floor. Amber tossed the piece of cotton over Kammie's head playfully.

Kammie snatched it off, wadded it up, and tossed it back. "Well, good."

"This one looks like a skirt of bananas!" Wren pointed to a scribbly drawing.

"Oh, yeah, that one didn't really work."

"You know what I was reading last night in that spy book? The one about World War Two?" Wren began. "There was this famous, beautiful woman named Josephine Baker. She was a singer and dancer. She was a lot more popular in France than in the United States, where she was from, but she was famous all over the world. She used to dance around in skirts made out of real bananas!"

"No, I've never heard of her. That's crazy," laughed Amber, "and she was a spy?"

"Yeah! Apparently, a lot of people don't know it but while she traveled around Europe doing shows, she carried secret coded messages and information to allies," Wren said.

"I wonder if she used a cipher code wheel!" pondered Amber.

"That reminds me, we really need to refresh our stock," Kammie said. "Tell you what, guys, I'll just fold up and glue some more periscopes, and prep more cipher wheels. How many printouts do we have left?"

Ivy pulled the sheets of printed card stock from a pile on the shelf and started to count.

"Quite a few!" she answered, not bothering to count past ten. She handed them over to Kammie. "Are you sure? It doesn't seem like as much fun as designing new stuff."

Kammie shrugged. "Maybe not to you, but I like it. I get intimidated trying to make stuff from scratch, and sometimes I just like to follow patterns and instructions. It's kind of soothing. Zen."

Just then Ivy's phone buzzed. She glanced at it, then grabbed Wren's arm and headed for the door.

"Have fun!" Ivy replied. "Mom's here to give us a lift. Wren and I have some shopping to do. We're going to make secret code invisible ink sets!"

MAKE YOUR OWN
BESPOKE UTILITY BELT

What kind of mission is it?

Undercover at a fancy party
Sneaking into an enemy lair
Reconnaissance in the jungle
(Your own idea)_____

~

What gadgets do you need?

Suction cups
Duct tape
Lasso
Periscope
(Your own ideas)_____
_____

What kind of fabric?

Fancy, elegant velvet
Strong, stiff canvas
Flexible, stretchy t-shirt fabric
Slippery, shiny satin
(Your own idea)_____

What else?

Does it need to be hidden?
Easy to get to?
How many pockets do you need?
Waterproof?
(Your own ideas)_____
_____

DESIGN IT:

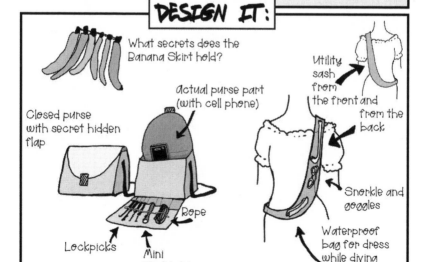

What secrets does the Banana Skirt hold?

Closed purse with secret hidden flap

Actual purse part (with cell phone)

Lockpicks

Mini Flashlight

Rope

Utility sash from the front and from the back

Snorkle and goggles

Waterproof bag for dress while diving

18

SECRET MESSAGE KIT TINKER

Wren stuck her hand into a bin of individually wrapped batteries. They were the big C and D ones.

"Hey, why are these wrapped?" she pondered. "What a waste of plastic!"

Apparently she said it out loud, because the clerk passing by said "If they weren't wrapped, their positive and negative sides would touch. The batteries would drain while they got hot, then they'd EXPLODE!"

The clerk spread out her hands in an imitation explosion. "Ka BOOM!"

"Okay," Wren replied doubtfully.

The clerk chuckled as she continued down the aisle of the electronics store and began hanging packages of conductive copper tape on a wall peg with a few similar packs.

Wren looked around for Ivy. She caught a glimpse of the tall girl's high black ponytail disappearing past the display down the next aisle. Wren ran around a rack of cell phones.

"I've been looking for you everywhere!" she said as she caught up with Ivy in an aisle of raw electrical components.

Ivy jumped, startled.

"Oh, sorry, I got distracted!" she said, motioning towards the contents of the aisle. Toggle switches, battery packs, buttons, capacitors, breadboards, and more hung in little packages and filled tiny labeled drawers all around them. "There are so many wonderful things here!"

"Explain to me again what we're looking for?" Wren asked, picking up a package containing a buzzer with wires dangling from it. The package proclaimed "SUPER LOUD!" and "PRE-WIRED" in bold orange capital letters.

Ivy took the package and hung it back in its place. "Not that. Something to use for secret messages. I thought we could make invisible ink kits. I watched a YouTube video about UV light a few months ago. It's a special kind of light that makes certain stuff glow in the dark. I have an idea if we can find the right components."

"Glow in the dark? What stuff? Like cats?"

Ivy paused and turned to Wren, totally confused. "Cats? What? No! Like fluorescent stuff."

"Fluorescent. Gotcha."

"Yeah, all sorts of stuff is naturally fluorescent, like scorpions, banana spots, and people's pee."

"Ewwwww!" squealed Wren, "Pee? That's gross! What else? Their poop?"

"I dunno, but a bunch of stuff. We can look it up later, but the point is you can only see it in the dark with special ultraviolet light. It's sort of purple. Some people call it blacklight. I think it's because it's not very bright? I don't know why they don't just call it purplelight. And they make special fluorescent ink we can hopefully find," Ivy stopped in front of a section labeled LEDs.

An assortment of plastic boxes sat on a few shallow shelves. Some were see-through and some were more opaque. Most of them were divided into compartments. Wren picked up one of the clear ones and took a closer look at the contents. Inside were tiny domes with two wire "legs" sticking out the bottom. One of the legs was noticeably shorter than the other.

"These are lights, right?" Wren asked, shaking the box.

"Yup," answered Ivy, pointing to the label of the package she held. It was larger than the one Wren had. "LEDs. Light Emitting Diodes. Makers use these to create these little glowing things called 'throwies.' You take one of those flat, round batteries that look like a fat quarter; a coin cell battery. The kind in car key fobs. Sandwich it between the leads, and the light lights up."

"And the leads are these two legs sticking out the bottom? The light up part is in the dome, then?" Wren was trying to wrap her head around it.

"Right! Exactly," Ivy nodded. "Then just stick a magnet on the side, maybe a little tape to hold the leads against the battery and throw it on a metal thing. That's why they call them 'throwies,' because you throw them. And then you have little fairy lights that stay lit until the battery runs out or the LED burns out. They're pretty cool. Mom brings them to parties."

"That sounds pretty," Wren imagined a bunch of colored lights stuck to the side of her mom's Prius, and looked more closely at the package in her hands. "But what does that have to do with ultraviolet light and glowing scorpions? And what does all this have to do with spy gear?"

Ivy held up a finger. "Because..." she ran the finger along a row of the little boxes until she finally found the one she'd been looking for. "...of this! All that fluorescent stuff will glow in the light from these. And trust me on the spy gear. I'm getting there. Seriously, it's a good idea."

Wren took the package Ivy held out to her and read the label. It said UV LEDs. Which Wren guessed was short for UltraViolet Light Emitting Doodads. "Ah HA! And these will light up scorpions in the dark?"

"Yup," Ivy continued. "When the leads touch the correct sides of the batteries, it completes the circuit and lights up the light. The thing about throwies, though, is that they're always on. What if we rigged it so you only touched the leads to the battery when you wanted it to light up? Like put something in there that was fatter than the battery. Something that held the leads apart so they didn't quite touch the battery on both sides unless you pressed it? Like cut a hole in a square of cardboard for the battery to sit in. The cardboard would hold the leads away from the battery unless you pushed them down to contact the battery's sides! The LED would light up like a flashlight."

"Okay, I totally understand," Wren nodded, totally not understanding. "But I might have to see it in action."

"Fair enough," replied Ivy, checking the quantity in the box. "Anyway, so then I thought, what if we glue a little loop to the cardboard and wrap the loop, the battery in it's cardboard square, and the leads of the LED in pretty ribbon? Leave just the light bulb part sticking out to shine on fluorescent stuff. Then just string a necklace cord through the loop and BAM! You have a fancy secret flashlight necklace to see invisible messages with."

Wren nodded enthusiastically. "Like beautiful light up ribbon mummy flashlights? I love it! But what are the spies supposed to write with? Don't say pee, because that would be disgusting."

Ivy rolled her eyes. "Of course not! We'll use this."

She held up a bottle labeled "Invisible UV Ink" that was sitting next to the UV LEDs. "Remember when Amber got that bulk pack of tiny empty lip gloss containers online back when she wanted to make her own lip gloss? We have like thirty of them still sitting around the Greenhouse. They have little roll-on applicator tops that can work like a ball point pen. We can fill those containers with the ink from this bottle and include it with the flashlights, and we have a spy product everybody will want. They write with the invisible ink, and read it with the UV flashlights! A kit!"

"We could hot glue a little loop to those too, and keep them together on the same necklace," Wren could already see the possibilities. She took the bottle from her friend and shook it vigorously as they headed to the counter.

As they walked past the bin of individually wrapped batteries, Wren asked curiously, "Hey Ivy, do batteries really explode if you put them together without plastic?"

"What?" Ivy laughed. "Who told you that?"

"The lady that works here."

"Ah! Yeah. She's funny," chuckled Ivy. "That's an exaggeration, I think. I've never heard about them actually exploding. But if you throw them all together, especially with metal or some other conductor, their positive and negative sides might connect and short circuit. Then they get drained and really hot, maybe even spark, and I guess they might start a fire maybe. At the very least they'd be useless. But I don't think they'd, like, explode."

"Huh," Wren shrugged. "I didn't know that."

"Yeah," Ivy rubbed the back of her neck. "I learned from the best teacher."

"Your mom?" Wren knew how smart Ivy's mom was.

"That would have been a lot easier," Ivy replied, "but no. Failure. Mistakes are the best teachers. I once threw a bunch of coin cell batteries in a ziplock and came back to a bunch of hot, dead batteries and a bag with little melty spots. Then I had to figure out why it happened."

The girls approached the checkout counter. Ivy's mom was leaning against it, chatting with the clerk Wren had talked to earlier.

"...and online shopping is totally killing us," the clerk was saying. "I hope we do well this Christmas."

"You'd think there would be a big demand for electronic parts in San Francisco, tech capital of the world," Ivy's mom replied.

"Oh, there is, but you can find all this stuff with the right search terms online. And I guess people don't like to get out much anymore."

"Or they're just too busy," agreed Ivy's mom. She suddenly saw the girls coming up with their stuff and moved to the side. "Find everything, ladies?"

Wren didn't answer. She was too busy glaring at the clerk, who didn't seem to notice.

"I think so," Ivy replied. "Except do you have any coin cell batteries? We need a bulk pack."

The clerk nodded and turned to grab some off the wall.

"Great! With those, I think we are all set to tinker up.

SECRET MESSAGE KIT

MATERIALS:

Coin cell battery
Corrugated cardboard
UV LED
Fancy tape
Pipecleaner
Necklace Cord
Cardstock
UV ink
Empty Lip Gloss container

Cut a 1.5 inch square of thick cardboard and 2 squares of cardstock.

Trace around the battery in the middle of the cardboard square. Cut out the circle and insert the battery.

Put the leads of the LED around the cardboard/battery. Pinch them against the sides of the battery. If it doesn't light up, flip the light around and try again.

Glue a pipe cleaner loop to the end where the necklace cord will go.

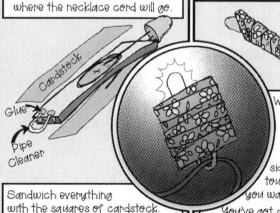

Cardstock

Glue

Pipe Cleaner

Sandwich everything with the squares of cardstock.

Wrap with fancy tape. String cord through the pipe cleaner loop and tie on the bottle of UV ink. Pinch the sides so the LED's leads touch the battery when you want it to light up. BAM! You've got a Secret Message Kit!

19

A CLUB DIVIDED

"Ivy, you've got to start taking the bus more or something," her mom said as they drove back to Wren's house. "Driving you around to all your games, practices, friends' houses, and now shopping trips is like a full time job! And I already have a full time job. Maybe we can find you a rideshare driver we trust or something. That would make my life so much easier."

Wren looked over at Ivy, shocked. Rideshare? Alone? But Ivy didn't look shocked at all. In fact, she looked proud. Wren knew there were special drivers for kids. And besides, Ivy was one of the oldest kids in sixth grade. Ivy wasn't the little kid Wren had met three years ago anymore. Sixth-grade Ivy was a tall twelve-year-old who acted like a teenager sometimes and had her own cell phone. Strong, mature, and confident, she sometimes seemed as out of place with the other sixth grade kids as

Wren. Just in a different, almost opposite way. Wren wondered if she'd ever feel confident enough to travel around the city alone, but Ivy seemed excited by the idea.

Eventually everyone went on busses and rideshares alone, right? Wren's mom didn't need an adult with her on the bus. When did it start? Was it in middle school? Would it be at a different time for Ivy than for Wren, or Amber, or Kammie? Ivy not having a dad meant her mom had to do everything, so maybe it wasn't right for Wren to compare herself. Before she met Ivy, Wren hadn't even thought about what it would be like to have only a mom. Not even a dad who lived somewhere else but you still knew was there. At least Ivy's grandparents were close enough to visit. They lived an hour away in Fremont, in the same house they had moved into when they came to California from Korea a gazillion years ago, before her mom was even born. Which was close enough to come to birthdays but still too far away to take Ivy to the store. Ivy didn't seem to be missing out on anything, but Wren had never really thought about how differently she had to do the little things Wren took for granted.

As they pulled into Wren's driveway, Ivy leaned into the front seat and kissed her mom's cheek. "Thanks a ton! I really appreciate it!"

Wren slammed the car door and waved goodbye to Ivy's mom.

"Wait, want to sleep over tonight?" Wren asked, afraid

Ivy would say she was too old now for baby things like sleepovers.

"Yeah!" Ivy nodded enthusiastically. Waving for her mom to stop, she ran to ask her through the car window.

Wren smiled with relief. Ivy might be older, and do things differently, but there was still a lot they could do together. She hoped the rest of middle school wouldn't pull them too far apart.

As the girls approached the Greenhouse chatting happily, they could see Kammie assembling gadgets at the card table amid a large pile of completed ones.

Amber sat at the potting table. But the sewing machine, light still on, was empty. Instead, Amber sat in front of an unrolled sheet of poster board. She was attacking it vigorously with a pencil and ruler.

Sticking her head in the door, Ivy asked a little uncomfortably, "We're back! What's up here? Everything go okay?"

Kammie looked up, then looked back down at her projects without a word.

Amber leaned back and sized up the word she was writing.

"I think so. I asked Kammie to brainstorm a slogan with me, but she wasn't very helpful. What do you think?" she said towards the poster board.

Ivy and Wren entered cautiously. Towering over

Amber's shoulder, Ivy read *Vote for Amber, Glow With Pride.*

Wren threw Ivy a glance and coughed. She turned her back to put the bag filled with their purchases next to the abandoned sewing machine.

Ivy shrugged, "That's... interesting. What does it mean?"

Amber grinned up at her, "I got the idea because you were talking about invisible ink. See, a candidate needs a catchy slogan so voters remember them on election day!"

Ivy rubbed the back of her neck. She forced a smile and looked over at the sewing machine. "Weren't you going to, um, you know, make your bespoke utility belt stuff to sell next week?"

Amber's grin disappeared. "You sound like Kammie. I made two of them to sell, plus a sample. I'll make more later. I've got to get these posters done for next week too. I'm way behind the other candidates!"

She looked at the others.

"Who's going to help me?"

Kammie said nothing. She pretended not to hear. Wren and Ivy looked at each other again.

"We were going to assemble the invisible ink kits. You know, like we talked about?" Wren smiled a not very convincing smile with a lot of teeth. "Gail is expecting us to have an expanded product line next week. Aren't you going to help?"

Amber rolled her eyes. "I already helped! I worked really hard on those utility belts. Take a look! Now I have other stuff to do! Get off my back."

Wren opened her mouth, but closed it again and turned to the pile next to the sewing machine. The "bespoke" spyware did look nice, even though it was pretty basic. There was a green canvas belt with a few straps and pockets on it, and the sash made from the green satin. Next to them sat a simple rectangle of blue cotton fabric with another piece of the blue fabric sewn down the middle. There were a few bits of velcro stapled to the sides and a pretty black ribbon strap sewn in. It looked unfinished.

Wren moved the fabric creations into the box where they kept the stuff to sell. She silently held the box open as Kammie dumped in her finished gear. Kammie blinked at Wren a few times silently, before turning back to glue the last mirror in her last periscope.

Dumping the finished product into Wren's box, Kammie said simply, "I have to go now."

She grabbed her bag and headed to the back door of Wren's house to call for her ride home. She didn't even look back.

"Is Kammie okay?" Wren asked Amber, who didn't seem to notice that her friend had left.

Amber made a raspberry noise and kept writing and decorating the poster with sparkling stickers.

No one said another word until Amber left.

20

BESPOKE

"**I**t's Amber!" Wren's mom said, holding out her cell phone to Wren.

Wren took it nervously. "Hello?"

"Hey, Wren," Amber said brightly. "Whatcha doing?"

"I was going to head to the art store to pick up more mini-mirrors for the periscopes," she replied, relieved Amber was in a good mood. "They're our best seller. We're almost out of supplies."

Amber sighed. "Of course. Well, I was going to head to Bespoke today to shop for an outfit for my inauguration. Wanna come with?"

"Your what? Where?" Wren suddenly remembered the fancy clothing store Amber told her about. "OH! Oh yeah. Is there even an inauguration for student government?"

"Dunno," Amber sounded a little grumpy all of a sudden. "But you know, just in case. Come with me. You'll love it! Besides, it's right by the art store, so we can do both. You know, together! We're leaving in a few, mom can swing by and get you on the way."

"Yeah," Wren smiled at that. "Yeah, I'd love to hang out. Good idea. Sounds fun. See you soon!"

Hanging up, Wren tossed her mom's phone on the couch and ran to get a jacket. "Going out with Amber, Mom! See you soon!"

Trixie looked up from where she was coloring. "I wanna go too!"

Wren tousled her hair. "Sorry sprout. Big kids only today."

Bespoke glittered. Literally.

Amber's mom held a sequin soaked t-shirt up against her daughter. "Oh isn't this precious!" she sighed lovingly.

Amber was running her hands over a pair of overalls. They were the same auburn color as her hair and seemed to be made out of velvet or something. Wren petted them to be sure. Velvet overalls? But Amber squealed happily and thrust them at her mom.

"I NEED these!" Amber gushed.

"Oh, you totally do," said a familiar voice.

The girls turned to see Axel standing behind them, holding a blue satin blouse and smiling at Amber. She held out the blouse and Amber ran her hands over it appreciatively. Wren touched it gently. It was beautiful.

Axel, with a barely perceptible nod at Wren, grabbed Amber's arm and started babbling about the auburn overalls and what shirt to wear with them.

Wren stood next to them, trying not to bump into anything. Then she noticed a dress that looked just her size. It was a mulberry color that she just knew would look great with her olive skin and blue eyes. The dress was fitted at the waist and flared out into a swooshing skirt. Holding it in front of her and looking in a mirror, Wren felt the unusual sensation of actually feeling kind of pretty. A handwritten tag swung from a bit of satin thread safetypinned to the care label. Wren gasped.

"Who spends this much money just on clothes?!?" she exclaimed, staring wide-eyed at the price tag.

"Apparently not you," Axel replied.

Amber giggled and Wren shot her a withering look.

"Sorry," Amber shrugged with an apologetic smile. "It was a little funny."

"Funny to who?" Wren frowned.

"Oh, don't be so sensitive," Axel rolled her eyes. "It was just a joke."

Wren just glared at her. "Come on, Amber, let's head to the art store."

Amber hesitated, a blouse still in her hands, as Wren began stomping away. She stood still by the rack next to

Axel, watching as Wren tugged open the massive door. Wren turned back to her.

"Come on. We can come back for whatever fancy clothes you want after we get more periscope mirrors. Let's get the important stuff out of the way and we'll have more time to mess around."

Amber stiffened, "Actually, this is important to me. I was hoping you'd help. You go ahead and I'll catch up later."

Wren looked around the glittering store, with its coordinated color palette and soft textures. At the styled mannequins and well-dressed customers. She forced herself not to look down at her sweatshirt, the one with the previous owner's name Sharpie'd on the label, and her faded jeans. She ran a hand through her unruly hair as a breeze from the open door blew in. She looked at Amber helplessly and mumbled, "How could I possibly help you?"

But Amber had already turned back to the rack and was pulling out a jacket to show Axel. As the door closed behind Wren, it cut off the high-pitched squeal from the other girls.

Wren stood on the sidewalk for a moment, disoriented by the noises from the street and overly bright daylight. She quickly regained her bearings and, lifting her head high, shoved her hands into the paint-stained pockets of her jeans and headed to the art store.

"You and your little club are getting pretty popular," Axel said in an offhand way as she tried on the jacket Amber had found.

Amber shrugged. "You think so? I dunno, to be honest, I'm getting a little bored."

Smirking, Axel handed the jacket back to Amber, who tried it on herself. "Aww. Poor thing. Written up in the school paper? Interviewed by Gail no less. Your little business is quite the buzz around school. Benjamin Spencer himself sings your praises..."

Amber's eyes shot up. "He does? Mine?"

Axel waved away the question without answering, "You can't buy that kind of name recognition. Seriously. My column in the paper doesn't even make me that well known. You're a shoo-in for a seat on the council."

Amber set down the headband she had been looking at and gazed across the store at her mom, who was carrying an armful of clothes into the changing room. She glanced over to the door. Wren had disappeared, probably to the art store. This wasn't how she'd thought the day would go.

"I hope so," Amber said quietly. "I think."

Tilting her head curiously, Axel asked, "Trouble in paradise?"

"It's just...I don't feel popular. I feel kind of lonely. I mean, making the spy stuff is fun and all, but that's the only thing my friends want to talk about. Nobody wants to do the stuff I want to do. Which kind of makes sense. I mean THEY aren't running for council."

Axel snorted, "They could at least be supportive. I mean jeez, what kind of friends are they? Sounds kind of selfish if you ask me. Student government is really important."

"Right?" Amber nodded at Axel, hands on her hips, "That's what I've been saying! They could at least support me. We probably have enough money to fix our microscope by now."

Seeing that all the changing rooms were full, Axel tossed her pile of clothes on a little table and flopped into a plush chair nearby. Reaching up to tighten her blond pony-tail, she shook her head in sympathy. "Well, they don't sound like very good friends to me. Sounds like they need their egos to be taken down a notch or two."

Amber turned on her. "Hey now. That's not fair. We don't always want to do the same things, but they're my best friends. There's nothing wrong with being excited about a business, I'm glad it's doing well. I've just got other things on my mind. It's not their fault. When I'm elected, they'll totally get behind me."

"If you say so," Axel picked up a magazine and started flipping through, waiting for an open changing room. "Certainly doesn't hurt your campaign to hang out with them. But I'd start looking for some new friends if I were you."

21

BEGUILED

*M*onday morning actually didn't suck. Ever since Gail's story ran on the second page of the newspaper, mornings had become easier for Wren. The election was still crazy and the kids were generally obnoxious, but lately instead of the eye rolls and kids whispering about what she assumed was her, Wren was greeted with a lot more hellos. Kids she didn't even know called out to her by name. Kids talked TO her instead of ABOUT her, which she couldn't remember ever happening before.

She wasn't really looking for their approval, but she had to admit, it was nice. And it had been her longest stretch away from Ms. Sophie's office since she'd started at Lovelace. With all the purposeful inventing going on, her brain didn't seem to have as much time to cause trouble. And she wasn't even bored!

As she approached her locker with the week's supply of periscopes, ciphers, and the new gear, the other sixth graders moved to the side to let her through. One of them even gave her a friendly nod.

It didn't take long for the new spy gear to sell. Within a week, they were almost out of periscopes again, and Wren felt extra smart about getting the new mirrors so they could restock.

Amber had even brought in a bunch of fashionable utility belt creations that she had made at home with the help of her mom. She'd made a special one for each of the other Renegades too, with their names stitched on them. They all looked really nice.

Kammie loved how Amber could design and sew, even without a pattern. She ran her finger along the dark teal K sewn on the simple gray satin armband pocket Amber presented her with. Understated, useful, and practical but still classy. She loved it. Amber really understood her.

Kammie would never have been able to design all the unique utility belts Amber dumped in the box, much less know how to sew them. She sold the first one to a young girl just as they were packing up after lunch.

"Enjoy it!" Kammie called after the girl, uncharacteristically loudly.

"You know, I don't think I've ever heard you talk before," said a strong voice from the other side of the table. "You have a nice voice. Do you sing?"

Kammie turned to see Benjamin Spencer standing just a few feet away. He was speaking to Kammie, but Amber suddenly appeared next to him.

"What a pleasure to see you again," Amber said pleasantly, looking up at Benjamin with a warm smile, the hint of a blush dusting her cheeks beneath her freckles. "Thanks for sending Gail to do that story on us. You were right about getting the word out. Business has been booming."

Benjamin held up his palms. "Nope, I didn't send her. That was her idea entirely. I did give her one of the cipher wheels, though. So for what that's worth, you're welcome."

"Oh, and..." Amber bit her bottom lip. "I decided to take your advice and run for student council."

Benjamin raised his eyebrows and nodded approvingly. "Oh? That's great. Good luck, we could use someone smart like you. So look, this isn't entirely a social call. I have a proposition for you."

"For me?" Amber squeaked.

Benjamin smiled and winked at her, "Well, for all of you."

"Yeah?" Wren asked, noticing Benjamin now that the gear and money box had been packed away. "We're all ears."

Sitting against the corner of the table, Benjamin crossed his arms. "It's like this, see. You know how the lockers don't lock, right?"

The girls nodded.

"So I'm on the swim team..."

"You're the captain of the swim team," corrected Amber. She quickly cleared her throat and followed it up with, "Or something. Right?"

"Something like that, yeah. I am," he agreed. "Anyway, sometimes, with the council, I have important things like paperwork. And I can't really keep anything safe while I'm in the pool... So, with the elections coming up, I'm going to have to take extra good care of the election results between when we count them at the end of the day and when I deliver the results to Ms. Sophie the next morning. But I have a swim meet that night here at our pool."

The girls were following along, but Wren's attention drifted to the other kids starting to trickle back into the school. Lunch was almost over, and Benjamin talked a lot. She hoped he'd get to the point soon. Amber nudged her with an elbow.

"So anyway, I thought maybe the Renegades could make me something I could keep stuff safe in."

"Like a safe?" Wren asked. "You want us to make you a safe to keep stuff safe?"

"Exactly," he pointed at her with a wink.

"Ummm. Why don't you just buy one at a store? We have one my parents keep our passports in. It's even fire-proof," Wren pointed out, ignoring the elbow Amber jabbed harder into her ribs. She pushed Amber's arm away.

"Well, yeah, I tried that," Benjamin admitted. "But they were all REALLY heavy. I need something I can carry around. And they didn't fit on the locker shelf anyway. Even with the smallest one, the door kept popping open."

Ivy nodded in commiseration. Just last week her locker had popped open and spewed her books and lunch into the hallway. She'd barely saved her sandwich from being squished by a candidate who was walking backwards, chatting with a voter.

"Okay, so why don't you just get a lock for, like, a duffel bag or something?" Wren persisted.

"Honestly, I've been searching for a solution for a while now. I've tried everything I can think of." Benjamin sighed and held up his palms helplessly. "Padlocks have keys I can't keep track of while I'm in the pool, those detachable combination gym locks banged around in my gym bag, and papers got torn and wrinkled when I tried the duffel with a lock. The perfect safe would be hard-sided, fit the shelf dimensions exactly, and have the lock built right in with no keys to lose or anything. But I'm not really a designer or maker. So I thought, why not get it custom made by someone better at making stuff than I am? Then I met you guys and saw all the great stuff you've come up with. I can't think of anyone else clever enough to pull it off. So here I am, asking. Begging. So, what do you think?" He clutched his hands together and gave them all puppy-eyes.

"Great!" blurted Amber. "No problem!"

Kammie sent her a worried look and Wren rolled her eyes, but Amber didn't notice. Her gaze never wavered from Benjamin's face.

"I don't know," Kammie said quietly.

"We did have the idea on our board but we hadn't really gotten around to it yet," Deep in thought, Ivy tapped her lips with steepled fingers. She looked like she was doing some calculations in her head. Focusing back on Benjamin, she asked, "You'd need this before the election?"

Wren fidgeted restlessly, watching as the stream of kids heading back into the school grew larger. Milo passed by and waved at her, pointing to the doors.

"It would sure help me out. I mean, it's not like anything would happen to the election results between the election night and the next morning, but I need a safe anyway, and that seems like a great first use, don't you think?" Benjamin's posture was relaxed, like he had all the time in the world. But his face looked earnest. "Please?"

"We've got to go in," replied Wren, hardly listening anymore. She kept staring at the doors. "Whatever, but I can't be late again."

"Is that a yes?" he looked at Amber.

"Yes!" she nodded, then looked at Ivy desperately. "Yes?"

Ivy shrugged. Wren started for the school, and Kammie followed, biting her fingernail.

"We can give it a try, I guess," Ivy agreed.

"Thanks! You're the best!" Benjamin called as he ran off to the school as well, easily overtaking Wren and Kammie, and disappearing into the doorway.

"I'm the best," Amber repeated dreamily.

22

BEFUDDLED

In the cluttered Greenhouse, surrounded by her best friends, with the warm sun streaming over her, Amber was bored. And she felt guilty about it.

She sat at the potting table making fliers. She had no interest in building the safe, but didn't want to leave the others hanging, especially since she was kind of the one who had gotten them into making this crazy thing. Plus, she just wanted to be with her friends.

Quietly, she opened her notebook to her latest fashion design, an outfit she might try to make when she won the election. On the next page, a list of homework and campaign tasks stared up at her. Very few were checked off except for some homework. Her eyes darted to the others.

Across the table, Wren was drawing spirals. Kammie was trying to balance a marker on her nose, and Ivy was looking at her watch, taping a pencil impatiently.

Amber's eyes drifted to the forgotten fliers in front of Wren.

"See how easy that was?" Wren nodded triumphantly. "We've got this. All we have to do now is brainstorm ideas, build them, test them out, figure out where they don't work, brainstorm and build again until we're happy with the result. That's how engineers get past being stuck! BAM!"

"What are we even brainstorming, Wren?" Amber snipped. "We already know we're making a safe."

Wren paused, looking a bit deflated. "That's true. Hmm. Well, do we brainstorm how to make it then?"

"Which part?" Kammie asked.

No one answered.

"Well," scratching her head, Wren looked at the engineering design process in her hand. "I mean, yeah. I guess we're past brainstorming and on to CREATING. Okay."

"So how do we create?" Ivy said. "That's the real problem I guess."

Wren sat down, looking confused. "I guess there's more to it than just following the steps."

Kammie's gaze fell on a printout of some computer code she was trying to write for a class. She'd been checking the print-outs for bugs on her way over, taking it line by line. Which suddenly gave her an idea.

"Maybe we need to break this big problem into small

steps," Kammie flopped the printouts at them. "That's what I do when I'm stuck on some nasty code."

"You're right," Wren agreed. "Let's just keep breaking it down smaller and smaller until we find a problem we can solve. Let's see. You're right that we'll have to make it out of something hard, but it has to have a door that moves in a material that doesn't bend. Oh I know! Hinges, of course. That's exactly what they're made for."

"That's something. A start anyway," said Kammie, perking up a bit. "I never really thought about hinges. I'm not even sure I know how they work. Which is weird since they're everywhere."

Wren motioned excitedly towards the door of the Greenhouse. "Right? There's all these amazing inventions we use everyday that we never even think about. Like zippers!"

She hopped off her stool and knocked her notebook to the ground but just stepped over it on her way to the door. She opened and closed the door a few times, pointing to the hinges. "I mean, how many doors do we all open every day? But have you ever really even looked at them? We can just buy a few little metal hinges I guess, they're pretty cheap and easy to find at any hardware store. We don't need to make them... althooooough... it wouldn't be too hard. Maybe we should try! We can use straws and skewers..."

Ivy interrupted her gently, "Nah, let's just buy them.

This project is hard enough. You can make your own hinges another time."

Wren shook herself to refocus and nodded.

"I agree," Kammie nodded, "let's buy them. Then we just have to figure out the lock. And the box material."

Amber looked up from the campaign locker magnets she was decorating. She pushed the magnetic sheets and glitter glue to the side and leaned forward. "Benjamin wants something like a combination lock, right? So, shouldn't we figure out how a combination lock works?"

Ivy scratched the back of her neck. "Well yeah, but we don't exactly have one around here we can take apart, do we?"

"No, but," Wren held up a finger, then turned suddenly and ran out the door, calling over her shoulder. "We do have the magic of the internet!"

Speechless, they watched her run into her house. They'd barely had time to look at each other in confusion before she re-emerged typing laboriously on an iPad with one finger, stepping deftly over a Trixie-sized shoe in the yard without even looking at it on her way back.

They all crowded around the iPad as the internet told them that safes were usually made of several layers of strong, heavy metal and that combination locks consisted of an alarmingly complicated array of tumblers, drive pins, wheel notches, drive cams, and a spinning combination dial. Ivy and Amber looked at each other in confusion and

Kammie started to breathe faster with every video they watched.

Amber sighed and, scowling, turned back to her locker magnets.

"How are we ever going to make one of those?" Kammie whispered, her hand over her mouth. "I don't even know what half of those words mean!"

Wren, still fixated on the latest video, wasn't listening as she hit play for the fourth time. A smile kept flitting across her rapt face.

"Look!" she said excitedly, pointing to the animation of tumblers falling into place as she watched it for the fifth time. "It's so beautiful! I had no idea locks worked like that! Look at how elegantly all the parts work together."

Ivy waved a hand between Wren's face and the screen impatiently. "Let's move on, okay?"

Wren flipped the iPad face down. "Okay, baby steps... we're going to buy hinges for the door and will need to make the safe out of metal, yeah?"

"But, how do we make something out of metal?" Kammie interrupted. "We can't cut it, fold it, tape it, glue it, or poke holes in it."

"Well, you can melt it," Wren suggested.

Kammie facepalmed and shook her head. "Do you know what kind of HEAT it takes to melt metal?"

Wren's eyes sparkled, "Oh, heck yeah I do."

They all thought for a minute.

"Hey, why don't we make our prototype out of cardboard?" Wren said suddenly. "Then my mom can weld it out of metal."

"Oh right," remembered Ivy, "Isn't your mom doing that welding class?"

Hope blossomed in Kammie's eyes, "Do you think she will?"

"Sure, why not?" Wren shrugged. "She's always looking for projects to practice on during her studio time. We'd have to pay for the materials, I'm sure, but I can't imagine why she wouldn't help us. But we do have to make a model for her to work from. A working prototype. Do we have any cardboard left?"

Ivy, Wren, and Kammie began to dig through bins and under piles of random paper for pieces of cardboard. It took a few minutes, but they eventually managed to find a pretty good-sized pile of scraps.

Amber, grumbling, found a piece or two. Then she suddenly stopped looking, and turned urgently back to her notebook, doodling furiously.

Wren grabbed a large piece of cardboard and cut up along the corrugation about six inches, and then over about four. She held it up, opening and closing the cut out.

"Here's our door," she said. "Let's pretend it's rotating on hinges."

Ivy was cutting another piece into a circle. "Here's our combo dial. Put it on the outside of the door."

Wren rolled up a bit of masking tape and stuck the disk to the middle of the door. She opened and closed the door again, swinging it in and out like an Old West saloon.

"Wait!" Kammie reached excitedly for the prototype. She re-stuck the circle further towards the edge of the door, so it overlapped the wall part of the cardboard. "Now the door only opens one way. The dial blocks it from swinging in. Or, it would if it was attached with something other than tape. If we put another circle on the other side too, the dials themselves will work like a lock!"

Amber, at the other table, pulled out a piece of poster board. She took out a pencil and started carefully copying something onto it from her notebook.

"Amber, can you hand me that cardboard next to you?" Kammie asked.

Amber looked up. "What?"

Kammie motioned to a scrap on the table. "That piece. Hand it to me. That one there."

Amber passed it over and turned back to her poster board.

Kammie cut the scrap into a circle the same size as the first one and taped it to the other side of the cardboard door. Then she secured both dials down tightly with a few strips of masking tape. The door didn't open, alright. In fact, it didn't go anywhere. The dials sandwiched the wall. The door locked as she'd hoped, but now it couldn't unlock. Kammie shook the door angrily. "There has got to be a way to lock it AND open it. But how? I suck at this! I'm so stupid! I just don't have a mechanical mind. Why am I so stupid?!"

Angrily wiggling the door, she tried to force it open while holding the cardboard circles in place at the same time.

Ivy blinked at Kammie not sure what to say, caught by surprise by the sudden wave of emotion.

"Can't I contribute ANYTHING?" Kammie whispered under her breath. With a final yank the cardboard door opened, bending one of the circles and tearing the door.

"Hey!" cried Wren angrily. "Don't do that! Now we have to start over. I think that was our last big piece of cardboard!"

And just like that, they were back at square one.

23

RAGEQUIT

*K*ammie stared at the pile of mangled cardboard. Getting so emotional over something as stupid as a bad idea was embarrassing, but she couldn't help it. She felt like a failure. Only, it actually hadn't seemed like a bad idea. Even now, with the destroyed cardboard in front of her and her friends looking at her like she'd grown a third head, using the dials themselves to lock the door seemed like a great idea.

With a deep breath, she picked up the largest piece of cardboard she could find and cut another sample door, holding it up hesitantly.

"Okay, break it down into small steps. Again. We have a door. We want to keep it from opening unless..."

"We KNOW that," snarked Wren irritably, whizzing a piece of scrap cardboard Frisbee-style across the Greenhouse.

The cardboard soared across the space and smacked Amber on the back of the head.

"OW!" Amber yelled, louder than necessary, looking up from the poster she had been carefully lettering.

The poster was almost done, "Vote for Amber, she's a Jewel!" written lightly in pencil. Amber had been meticulously tracing the carefully drawn letters with permanent marker. She was at the J when Wren's cardboard smacked her.

The surprise caused her hand to jump, the marker skittered up a few inches, leaving a trail of permanent ink behind. Like a scar across the perfect poster board.

Everyone stared at it. Amber began to tremble. Her face flushed a bright red.

She lifted fiery, blazing eyes to Wren screaming, "You are SO SELFISH! Look what you DID!!!"

"I'm so, so sorry, Amber," Wren cowered. "Really! I didn't mean it. It was an accident. Let me help you fix it. Maybe we can make that line into a flower or something..." but her voice trailed off when she looked up into Amber's face. "Or... maybe not?"

Amber shook and her eyes glistened with tears. She didn't answer. Instead she grabbed her backpack and violently shoved stuff in. Her notebook, posters, and the stack of magnets she'd been drawing on with glitter glue. The glue smeared all over her fingers.

"Calm down Amber, she didn't mean..." Ivy tried.

Amber spun on her friend. "You!" she pointed. "Just shut up!" She glared wildly at each one of them, "I'm trying to run a campaign! None of you even care! What have ANY of you done to help me? I just want to make a difference at school!"

Wren snorted and snipped back, "You aren't running because you care about the school."

Amber ignored her. "I just want something for ME and you're supposed to be my friends. Why aren't you helping me? You're supposed to CARE but you haven't helped me with a single thing! Even Axel's friends are helping her and I don't have ANYONE! You're all too wrapped up in this stupid spy stuff to think of anyone else. When was the last time any of you even THOUGHT about the microscope?"

She breathed heavily, jabbing a finger at the dusty cover of their precious piece of equipment. She grabbed her jacket and stuffed it into her backpack on top of the magnets, glitter glue smearing over the soft white cotton.

Kammie looked at the floor, hands holding the cardboard door hanging limply at her sides.

"Look, Amber, we're sorry, but the spy stuff is our business now. We're making money we need for the club," Ivy's face grew stern. "And we made a promise to invent this safe. You made a promise. It's important to keep your promises. You surprised us with the whole council thing. We didn't plan for it, we didn't discuss it."

It was like reasoning with a tiger on a rampage.

"So that makes it okay? Because I didn't ask for your permission first? You're not the boss of me, Ivy. I guess you guys have your priorities and I'm not one of them," she flipped her auburn hair in Ivy's direction as she reached for the door. "This safe is pointless and we're never going to figure it out. And if all we're ever going to do again is make more stupid money, I QUIT!"

Wren's mouth had been scrunching up tighter and smaller, her eyes narrowing. But she hadn't said anything until Amber's final two words.

"Excuse me?" Wren's voice came out low, slow, and ominous. "You want to quit, you go right ahead. You haven't been helping AT ALL anyway." Her eyes blazed and her voice got louder. "Why are you even here? To work on your stupid posters while we do all the real work? You decide to run for student council without even telling any of us, then expect us to drop everything to help you what? Get popular? Spend more time with Benjamin? Do you think he even cares? You couldn't care less about making a difference at school, I mean come on. That's not your team, Amber. You're supposed to be part of this team. THIS team. But you've been totally wrapped up in your own stuff for weeks now. And you're the one who made this stupid promise to Benjamin in the first place, and what? Now it's too haaard? Now you're leaving US to clean up your mess? YOU'RE the selfish one!"

Amber stared at her, breathing hard, one hand on the

Greenhouse door, the other holding her backpack. With a growl, she stomped across the yard towards the main house, kicking Trixie's shoe as she passed.

No one said anything. Ivy looked from Wren's angry face to Amber's retreating back, and turned away from them both.

Wren started picking up and slamming down cardboard squares. When she looked towards the sliding glass door into her kitchen, Wren could see her mom's face looking out towards them. Her mom's hand was around Amber's shoulders, a phone in the other hand. Amber was not looking back.

Wren turned away and said to no one in particular, "You just quit then. We don't need you anyway."

24

SUCCESS MINUS ONE

The Greenhouse was somber the next day. The emergency Sunday meeting had been planned since Friday, but Amber didn't show up. Wren sat at the potting table across from an empty stool. The two sad little stickers she'd stuck to the "Vote Amber" flier looked up at her like accusing eyes. She tried not to imagine they were scolding her, and failed. Failed.

Just like she'd failed with Amber, and failed with this stupid safe.

Wren swept the entire stack of fliers into the recycle bin.

"It wouldn't have been that hard to put on a few stickers, really," Wren sighed to Kammie, who sat at the card table with a cardboard "door" and some circles. "Then we could have gotten it out of the way together and moved on."

Kammie looked at her, sniffed, and turned back to her

cardboard. She had tried to glue a circle to the door before realizing it wouldn't spin after it was glued.

Wren wasn't used to seeing Kammie persist. She usually gave up when things got hard. Despite her mood, Wren found herself watching Kammie manipulate the cardboard. Maybe the safe wasn't a failure. Maybe it was just hard. And hadn't they been proving these last few weeks that they could do hard things? Failure would only happen if they gave up. She hadn't failed yet. None of them had, and here was Kammie of all people, keeping at it. If she could do it, they all could. This problem could still be solved. She just needed to rally the troops, the remaining ones anyway.

"Alright, Renegades!" she chirped. "What have we got? Let's review. What's our problem?"

"So, okay," Kammie stared at her cardboard door and dial. "We've got a dial we can use as a combination lock. If the dial is in the middle of the door, the door swings open both ways, in and out. If we put the dial on the edge of the door so it overlaps the wall, the door only opens one way. If we put a dial on BOTH sides, the door doesn't open at all. Any ideas on how to make it so it's only locked sometimes and not others?"

"Actually," Wren turned away from Amber's empty stool, "I couldn't sleep last night because my brain kept throwing out ideas about it. One of them was pretty good. What if we cut off part of the inside dial, right? Make it like a D shape. Then if the flat part is lined up with the edge of the door, it opens, but if it's rotated so any other part of the D overlaps the wall, it won't."

She reached for Kammie's door and dials. Cutting one into a D shape, she taped it in place. Sure enough, the door opened. When she rotated the dial and retaped it, overlapping the edge, the dial stopped the door from opening.

"Of COURSE!" Kammie's eyes grew wide. "It's so obvious now that you point it out!"

"We never would have gotten this far without your great idea, Kammie," Wren smiled at her. "Using the dials as locks was brilliant."

A smile bloomed on Kammie's face too.

Ivy took the cardboard prototype, opening and closing the door, spinning and retaping the dial in different positions.

"Okay," Ivy nodded. "So we just need to put this disk on an axle so it can spin. What can we use as an axle? Maybe a BBQ skewer?"

She handed the template mechanism back to Wren and dug through the bin marked KITCHEN STUFF.

Pulling out a long, thin, bamboo skewer, the kind used for kebabs, she poked it through where the center of the D-shaped dial would be if it were still a circle. Then, lining up the flat edge of the circle with the flat edge of the door, she poked the skewer through the door cardboard too. The dial spun on the skewer like a pinwheel. Ivy took the second circle from Kammie and popped it on the other side, sandwiching the door between them. The door could still open one way, when the D-shaped dial was in the right position. The full overlapping circle on the other side kept the door from opening the other way.

Wren watched as Kammie reached over and spun both dials in different directions on the skewer.

"Wait!" bubbled Wren, reaching for the mechanism. "Plug in the glue gun! I'm about to be brilliant!"

She glued both cardboard dials to the skewer with the door snug between them, but made sure not to get any glue on the door itself. Now the dials and skewer could be rotated together, as one piece, independent of the door. Rotating the full dial on one side rotated the D-shaped circle on the other. She could lock or unlock the door by rotating a dial.

"Coooooool!" said Kammie. "You're so smart!"

"Who's smart?" asked a voice from the doorway.

Trixie held a bowl full of cookies. Her elbows stuck out like chicken wings as she gripped a bunch of milkboxes under her arms. Ivy reached over and caught the milk boxes as Trixie raised her elbows. There were five of them.

"I carried them all the way here and didn't even drop one!" Trixie said proudly. "Hey, where's Amber?"

"Not here," replied Kammie curtly.

"Oh," Trixie pouted. "Boo. Whatcha doing?"

They showed her the lock. Trixie played with it enthusiastically while they watched.

"The problem is," Wren continued, telescoping a straw out to full length and stabbing her milk box with it, "it's really easy to crack. I mean, just turn and tug until the door opens. Not very secure."

"Yeah, a combination lock usually has three numbers. You turn it one way, then the other, then the first way again," sighed Kammie.

"But it's made of all those tumblers and things all falling into place and we can't possibly figure out how to make and build that before the election," said Wren. She took a long drink of her milk. "But I still think this idea will work."

"What about two ones?" Trixie suggested. "If you don't like one one, you can make two ones. Right? Two is my favorite number. And three is my favorite number, too!"

"Great idea. Two dials," Wren agreed. Trixie beamed, stuffing a cookie into her mouth. "The problem is rotating two dials. We could stack them along the door, side by side, but it would depend on how big the door is. If the safe fits on the locker shelf, the door can't be very tall, so we probably won't have room for more than one dial."

The others nodded. Trixie nodded too, even though she didn't have a locker yet.

"But," Wren continued, "if we put the dials on top of each other and used two skewers, you know for two dials, we'd have to poke into the other. Then it won't rotate anymore. And if we put both sets of dials on the SAME skewer axle, they'd just rotate with each other. We need them to rotate one at a time. Hmmmm."

Wren pulled the straw from her milk box and tapped the top. Kammie handed her a paper towel and pointed to the tiny drops of milk splattering everywhere, but Wren kept tapping. She took the towel and absentmindedly wiped the milk drops, then stuck the straw back in for another drink. Suddenly, she stopped sucking, lifting the box to stare at the straw.

Setting down the milkbox, Wren grabbed a bin and started digging through it, tossing out random pieces of string, wooden clothespins, and pipe cleaners haphazardly. She finally found what she was looking for; a plastic drinking straw. She held it up and looked through it.

The other girls watched, confused. Two cookies stuck out of Trixie's mouth while she tried to fit another one, dropping crumbs all over. Ivy moved the bowl out of reach, grabbing one for herself, and watched Wren curiously.

Wren grabbed several more pieces of cardboard and

handed them to Kammie. "Here, cut me another door and four circles, one circle needs to be smaller than the other three. I have an idea. But we need proof of concept."

"You got it!" Kammie started cutting. "What's proof of concept?"

"Like, making a prototype to prove an idea works" Wren said into the bin she was digging through. She pulled out a straw that was thinner than her first one. "Once we know the idea works, we can go from there."

She fit the thin straw inside the thicker one and nodded.

By the time Kammie finished cutting, Ivy knew what Wren was doing. She looked around until she found a wooden golf tee.

"I think this will be thick enough for the hole," Ivy said, handing over the tee.

Using the tee, Wren punched a hole through the center of all four circles. She lined one up so it overlapped the door, and used its hole to figure out where to punch the door's hole. Then she drew a line along the edge of the door right onto two of the larger circles and cut along it, making them into door locking Ds. Then she stacked the smaller dial on top of the larger one, matching the holes.

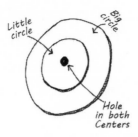

Next, Wren put the big dial on one side of the door, and a D on the other side, lining up all the holes. Then, she stuck the widest straw through everything.

She hot glued the big dial and the D to the straw with the door sandwiched snuggly between them, and trimmed the straw flush with the cardboard. Then she poked the thinner straw through the other D, glued it, and stuck the whole contraption into the wider straw. Last, she glued the smaller dial in place. Finally, she held the whole thing up and showed it to the others. The cardboard was stacked: small dial, large dial, door, D-shaped lock, and the other D-shaped lock, with one dial and lock set attached to each straw, and the two straws nesting inside each other.

Each dial turned one (and only one) of the locking Ds on the other side of the door. The thinner straw spun inside the thicker one. To unlock the door, both dials had to be in the right place.

"Ah," Kammie gasped, "I get it now. You have to line up BOTH of the dials. That's a lot harder than just one. Each dial you add multiplies how hard it is to guess the combination!"

Drawing a little arrow on the cardboard wall, Kammie dialed the locks so the door opened and wrote "o" on the dials where the arrow pointed. Quickly, in her neat handwriting, she filled in the rest of the dials with numbers, and handed it to Ivy.

Ivy spun the dials, turning the small one to o and the big one to 3. The door wouldn't open. She dialled both to o and the door swung open.

"Oh man," whispered Wren. "This is awesome. Trixie, go get Mom."

WREN's very impressed mom fiddled with the cardboard lock, "Yes, I'll need some help from my instructor, but I should be able to copy this prototype with metal. It'll be fun! Nice work, kids!"

They decided on three dial-and-lock sets, all nested inside each other.

"What about the combination?" Wren's mom asked. "If I etch the numbers right into the metal, your friend can never change the combination. So, then we'd all know the combination. Forever."

The girls hadn't considered giving Benjamin the ability

to change the combination, but it seemed like a smart precaution. Luckily, Wren's mom had a solution.

"What if you just use stickers? We have some at the shop that would be a good size. Your friend can use whatever number and letter combination he wants, and can take them off with a little rubbing alcohol."

"Sounds good to me, Mom!" Wren agreed.

"This may be one of the best inventions we've ever tinkered!" Ivy said. "I'm so impressed with us!"

"I just hope Benjamin likes it!" agreed Kammie.

CARDBOARD

MATERIALS:

Lots of corrugated Cardboard
1 Big straw and 1 smaller straw
that fits inside the other
a Hot Glue Gun is helpful
So are big scissors
an awl or golf tee to poke holes

Make a fake door out of cardboard.

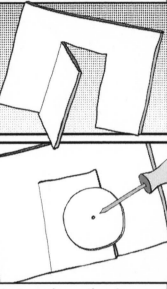

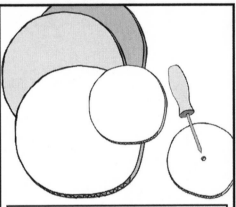

Cut 4 circles, 3 big and 1 small, from cardboard. Poke holes in the middle.

Put circle on the door so it overlaps the edge. Poke a hole in the door where the circle's hole is.

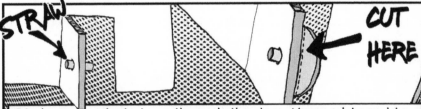

STRAW

CUT HERE

Stick your widest straw through the door. You might need to widen the hole, but try to make the hole snug to the straw. Add a big cardboard circle to the straw. Open the door and draw a line on the circle, then cut the circle so it doesn't block the door from opening. Cut another big circle to match.

COMBO LOCK

Put a full circle on the other side of the door.

Hot Glue both cardboard circles to the straw (but not the door). Trim the straw flush to the circles once the glue is cool.

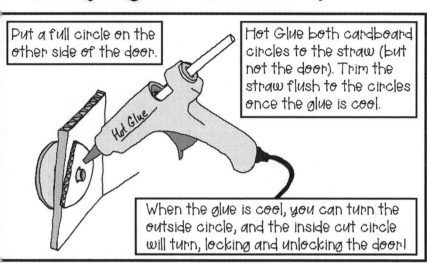

When the glue is cool, you can turn the outside circle, and the inside cut circle will turn, locking and unlocking the door!

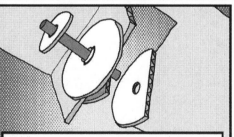

Add numbers, letters, or symbols, and line up your combination with a arrow on the front.

Add a second dial by sticking the thinner straw inside the wider one, adding the smaller circle to the front, the other cut circle to the back, and gluing them to the thin straw.

UNLOCK LOCK

When the front dials turn, the back dials will too. Your combination lines up the flat side of the cut circle with the door, letting it open.

25

TRIUMPH!

"*T*hat is the most perfect thing I have ever seen!" Benjamin gushed, twisting the dials of his new safe. "I KNEW you guys could do it!"

He elbowed Gail Mendez. They'd been chatting together when the three girls came to deliver the safe and a sheet of metal stickers.

"Did I tell you, or what? These girls can do anything. They're going to change the world."

Wren's mom had engraved little dots on each dial to show where it unlocked. Gail lined them up and opened the door investigating the elegant locking mechanism inside.

"Brilliant," she looked up at them. "Hey, can I do a feature on it for the front page of the paper?"

Ivy stood up straighter. "Really? For the front page?"

Kammie nudged Wren and jerked her head towards Gail.

"Oh! Yeah," Wren added. "But you should know we probably won't be taking any more special orders any time soon."

Kammie shook her head forcefully.

Benjamin chuckled and patted the safe, "It was that tough to make, huh? Well, I feel privileged. Thank you all." He suddenly looked around. "Hey, where's the other girl? Your red-headed friend?"

Ivy fidgeted and Kammie looked at her shoes.

"Amber is running for student council," Wren said diplomatically. "She's been busy."

"Yeah," Benjamin sympathetically. "That can take a lot of time. Well, tell her thanks and good luck from me next time you talk to her."

"Will do." Wren tried not to sigh. "Next time we see her."

For Wren, that would be the morning's science class. She walked in, early for once, not sure what she was going to say. But Amber was sitting with Bobby instead of at their usual table. Amber, avoiding eye contact, opened her notebook and started talking to him. Wren screwed up her mouth and plopped down in her regular seat.

Milo sauntered in. Seeing Amber sitting next to Bobby, he joined Wren instead.

"I read about you guys in the paper!" he held out his

hand to shake hers. "Congratulations on your business. Seems to be going well, huh?"

"Thanks," Wren replied half-heartedly. Celebrity had gotten a lot less fun without Amber.

Milo nudged Wren in the shoulder. "Save one of those periscopes for me, okay? I've been meaning to stop by at lunch. Maybe today."

"You bet!" Wren smiled as she pulled out her worksheet, "What did you get for number twelve?"

Out of the corner of her eye, she saw Amber glance over as she and Milo laughed together about his worksheet answer. Amber's face turned from wistful to sad, and then to angry.

26

A HOLLOW VICTORY

*G*ail's paper, *The Lovelace Gazette*, came out mainly online. Every Thursday, though, she distributed a few dozen print copies at lunch.

Kids swarmed around like vultures, waiting for the most recent edition. They squabbled, they talked with little bits of their lunches spraying out of their mouths, they smelled like tuna fish and peanut butter, and generally were normal kids eating lunch. *Disgusting*, Wren thought. She usually avoided the cafeteria.

No one but Wren ever seemed to notice how awful it was in there. Just more proof that her brain worked differently than theirs. Like how everyone else seemed like an emotionless robot and got so annoyed when she reacted with what, to her, seemed like regular emotions. They also apparently didn't hear or smell very well. How else could they stand it?

But today Wren stood pressed against the cafeteria wall among the chewers. She stood bravely, breathing evenly, waiting for Gail to bring out the newspaper. She had promised she'd get a copy since her locker was closest to the cafeteria, so she was going to make sure she got one before they were gone. No matter what. She'd stayed away from trouble, kids were being super nice to her, she could do this one little thing!

Gail arrived and Wren pushed her way through the crowd for a newspaper. Gail saw her and reached through the grabbing arms to press a copy directly into Wren's outstretched hand. As soon as Wren's fingers closed around it, she rushed out to the recess yard.

Breathing air blissfully free of other kids' lunches, Wren paused. Just for a minute or two as her senses calmed down.

Their usual picnic table waited for them, the other kids respectfully avoiding their spot in the mad rush for real estate. Wren saw Ivy and Kammie already setting up. A few kids waited as they pulled out their spy gadgets.

"Hey there," Milo appeared at Wren's side. "I made it. Let's go see what you guys have."

They headed over to the table together.

After Milo left with his new periscope and secret message kit, Wren took a look at the newspaper. As they'd hoped, Gail had already written the feature about

Benjamin's safe. It was on the front page along with a photo of the three Renegades, Benjamin, and the safe.

"THE RENEGADES CREATE ANOTHER WINNER FOR SPIES!" the headline read. The article explained how they created the safe for Benjamin to keep the official election results safe while he was at his swim meet, and included quotes from all three of them. It even mentioned Trixie by name. Amber only came up briefly at the end.

"Look guys!" Wren was giddy as she held up the paper. Ivy and Kammie crowded around to read it.

"Wow," was all Kammie could say.

And there it was, thought Wren. Proof of what she could do. Not by being boring, not by being bored, not even by being like the other kids. Wren read the praise-filled article again, hungrily. A warm glow filled her as pride spread almost physically through her body with each word. She'd worked through the hard stuff, and she'd been good at it. She hadn't been the one to quit, she thought with a pang of sadness. Wren shook it off. Amber had made her choice. Wren and the others had made another choice. They'd chosen not to quit. And together, she, Kammie, Ivy, and even Trixie had done something they'd thought was impossible at first, just by persisting. Maybe she had some hidden strengths after all.

"Nice work, Renegades!" Ivy held up both her hands for high-fives with a glowing smile.

Just then, a small group of kids approached their table and they went into sales mode. After that, kids lined up to get their hands on Renegade Girls' spy gear. It was hard to keep up with demand, especially since they were only a team of three now. Well, four if you counted Trixie, and Wren had to admit she kind of did count her now. Unfortunately, the younger grades had their own playground so Trixie couldn't join them. Usually Wren didn't think much about her sister at school, but right now they could have used her help with sales.

"Hey guys," called Wren as they were packing up their few remaining items that Friday. "Anybody seen the sample periscope? The purple one?"

Ducking under the table, Kammie called out, "Not down here." She looked around on the seats. "Did you pack it up already, maybe?"

A tall boy with long bleached hair handed Ivy exact change for the green canvas utility belt. Amber had been right. The specialty items were selling surprisingly well, despite their higher price. Kids seemed to like that they were one of a kind.

Ivy thanked him and put the money in their money box, closing the lid. "Nobody accidentally sold it, did they?"

Kammie and Wren shook their heads.

"I haven't sold a periscope all day," replied Wren.

"Me neither," added Ivy, "But yesterday my personalized utility belt disappeared too, and I still can't find it. Things are getting so crazy!"

Kammie walked all around the table searching the ground, but they hadn't dropped anything. She pulled on the ribbon that held the money box key around her neck so she could lock up. When she reached across the table for the box, she was surprised to see Axel looking back at her. Reaching for her book, she was surprised to see Axel looking back at her.

"Oh. Kammie. Umm, hi," Axel hugged her backpack and lifted her hand in a small wave. She looked slightly embarrassed. "I mean. Yeah. Hi."

"Hey," replied Kammie, but it was more of a question than a reply.

Axel didn't usually come over to the picnic tables during recess, and none of her friends were with her.

"Did you want something?" Kammie asked quietly.

"I... umm..." Axel fidgeted. She pointed to the UV flashlight kit that Kammie was putting away. "What's that?"

Kammie handed it to her.

Ivy looked up from searching through her backpack. "That's an Invisible Message Kit. You write with the UV ink but can't see it until you shine the flashlight on it." She demonstrated for Axel, who was uncharacteristically quiet. "Pretty cool, huh?"

Axel stared at the invisible message, shining the light back and forth, deep in thought. "Yeah. Actually. You know, this stuff really is awesome. I can see why you made the front page. I write for the paper and haven't ever made the front page. But you guys just..." she stopped herself and looked up at Ivy with pursed lips. "You guys did good. You really make cool stuff."

Ivy was surprised, and smiled warmly. "Thanks!"

"Did you want to buy that?" Wren asked Axel, pointing to the invisible ink kit. "We have to pack up now."

"Oh!" Axel blinked. "I mean—yes. I mean, that's actually why I'm here."

Kammie took her money and unlocked the money box. When she looked up again, the perky blonde ponytail was quickly disappearing into the crowd.

As they headed back to class, Wren saw Amber's back. She tried to call out to her, but Amber was too far ahead, chatting with Axel about something. Wren saw Amber angrily toss some crumpled paper into the recycle bin. Peeking into the bin as she passed, Wren saw it was a copy of *The Lovelace Gazette.*

27

REUNION?

For the first time ever, Wren didn't want to go to science class.

It was Monday, the day before the election, but she didn't care about that. She had even finished her homework for once, so that wasn't the problem either.

Yet, standing in the hallway outside the science room, she felt sick to her stomach. It wasn't fair. Things were going so well otherwise. She wanted to wake up and be friends with Amber again. But that wasn't magically happening. Amber hadn't shown up for the Renegade meeting on Saturday, and no one had even mentioned her except Trixie.

But Amber was on the other side of that door.

It had become so uncomfortable to be in that class that Wren almost pretended she'd thrown up that morning so she could stay home. Maybe it wasn't too late, she thought.

Maybe she could head to the office now and pretend she'd just thrown up. Maybe...

No.

That was no way to be a problem solver. She was stronger than that now. Wren took a deep breath and yanked open the door. Amber sat at her new table with Bobby. As soon as Wren made eye contact, Amber turned away to talk to him.

Wren flopped her books down at her usual seat and sat heavily.

"You really should talk to her," said Milo, who was already leafing through his homework. He didn't even look up.

"What do you mean?" asked Wren innocently.

Milo stopped leafing and looked at her, but didn't say anything.

Wren fidgeted under his gaze.

"Well, I mean, I've tried to but she's being stupid. Even Kammie and Ivy think so."

Milo raised his eyebrows and continued to look at her silently.

"Oh shut up," she mumbled, heaving herself up out of her chair. "Fine."

He smirked, watching her stride purposefully across the room.

Amber looked up as Wren approached. "What do you want?" she snipped.

Wren faltered. She stopped short and dug her hands into the back pockets of her jeans with an uncomfortable shrug.

"I just," she struggled. "You know."

"No, Wren. I don't know. Why don't you tell me?"

"The... well... the utility belts are selling well..." Wren began.

"Oh? That's nice. I suppose you expect me to make you more? Well, I'm too busy. I have other things going on in my life, you know."

Wren tried again. "That's not what... I just... they're really nice, okay? You were right about them."

Amber looked at Wren and her expression started to soften. Then her gaze fell on Milo, who watched Wren with a smile that looked almost proud, and her face got frosty again. "What, I'm not useless after all? I actually had a good idea about the bespoke gear? Too bad you were too much of an idiot to realize it when it actually mattered!"

Wren's mouth moved but no words came out. Finally, she managed to say, "No. I mean, we're doing just fine without you and your stupid buttspank stuff anyway."

"Be-spoke," Amber snipped as Wren turned to stomp off. "It's not that hard. Bespoke! Can't you remember a simple word?"

Wren looked back over her shoulder, abruptly aware that several other kids were watching, aware that class would start in a few minutes, but unable to let the comment go.

"You know what, smartypants?" Wren said crisply. "I looked it up. That's not even what bespoke means! It means 'made to order' not 'one of a kind.' So your utility belts aren't bespoke at all. The only really bespoke spy thing we've done is the safe. For your favorite Benjamin. Which you didn't even bother to help us make!"

Amber glared silently at her, tears glistening the bottom of her eyes, before spitting, "I don't care. You go bespoke all by yourself! I have better things to do with my time than spend it with people who don't even care if I'm around!"

Wren's head whirled and buzzed with anger, shame, snippy retorts, and confusion. Before she could say anything or even know what her body was doing, she found herself back at her seat.

As she sat down, hardly aware of anything except the raw emotion flooding her brain, she vaguely heard Milo say, "Sorry."

Now she really did think she might throw up.

28

ELECTION DAY

*I*t rained on election day.

By the time lunch recess rolled around, the rain had slowed to a drizzle just light enough for the kids to be forced outside, but just wet enough for no one to be happy about it. Candidates watched their last minute poster efforts get soggy, as the words fuzzed and ran down the poster boards.

Amber's lovely light pink sundress, chosen with such care the night before, was hidden beneath her paisley-patterned rain jacket. Auburn locks that weren't under her hood stuck unceremoniously to her face. She stood alone in a small puddle that sploshed up around the thick soles of her rain boots whenever she moved.

Holding a small pile of moist fliers, looking for someone to give them to, Amber was on the verge of tears.

She watched as a group of kids, including Kammie,

Ivy, and Wren, gathered under an overhang close by. Despite the misty drizzle, the kids were laughing and chatting brightly.

Amber saw Wren playfully punch Milo in the shoulder. He pretended to be injured, then said something and pulled her hood over her wet hair. She couldn't hear what they were saying.

She turned away. Bobby, Afsheen, Tyrone, and a few others she didn't know stood along a section of the school wall with her. They were only allowed to campaign in one area this late in the race. Voting would be held right after lunch. Axel was all the way at the other end, too preoccupied to even notice Amber's attempts to wave at her.

Amber caught Tyrone's eye and smiled at him. He smiled back and pointed helplessly at the rain with a shrug before facing forward again.

An unlucky student was making her way along the sidewalk, trying to get to the door of the school. The student, an upper grade girl dropped off at the end of the alleyway by her dad, had probably been at a dentist appointment or something. With a quick look at the line of candidates between her and the door, she took a deep breath and pulled her hoodie over her face. Quickly striding down the sidewalk, she was accosted by fliers, the last of someone's personalized locker magnets, and a ribbon on a safety pin, all bearing the name of some candidate or another.

Hands in her pockets, the girl refused to make eye contact with any of them. Amber let her pass without trying to hand her anything.

Lunch recess finally ended, and the kids filed back into the school. Barely used umbrellas tapped against the door frame and boots kicked the industrial sized door mats, spraying drops of water.

Amber morosely imagined her own face on the doormat with the slogan "Vote Amber, She's a Jewel" as she watched the feet stomp against it.

"Hey Amber! Good luck! I'm going to vote for you!"

The voice caught her by surprise. Milo smiled at her, both thumbs raised supportively. Then he was gone in the crowd.

Sitting alone in her afternoon class, waiting to check the box next to her own name on the long ballot, Amber was no longer sure she even wanted to win.

CONSIDERING the weeks of frenzied and chaotic campaigning, voting itself was kind of anticlimactic.

After lunch, the teachers distributed lengthy forms with a list of names and checkboxes to every student. Within ten minutes, they'd retrieved them again, marked with votes. The forms went into boxes, which were delivered to the art room.

Art classes had been canceled, and Benjamin Spencer and his four remaining council members were excused from their afternoon classes to tally the results. Counting would take until after school ended.

At that point, the winners would be known only to Benjamin and the others counting the ballots. As president, he'd hang on to the tally sheet to deliver the official copy to the school administrators the next morning. They would then certify the vote.

Since he had his swim meet that evening, Benjamin brought his fancy new Renegade spy safe, which fit neatly on the shelf of his locker. Of course no one expected any trouble, but it would keep the tally sheet and ballots secure overnight.

Gail would post the winners online in the newspaper by lunch the next day and distribute hastily printed special election edition copies in the cafeteria. If everything went well, there would be an assembly that afternoon for the new council members to be announced, cheered, and welcomed into the council.

Only, none of what was supposed to happen actually happened.

On Wednesday, the day after the election, no one had any news. The teachers didn't know anything, the website didn't get updated, and there were no papers brought to the cafeteria. Benjamin and the rest of the council were pulled from their respective morning classes. No one could find Gail Mendez either.

Amber and Axel sat together at lunch, but Amber couldn't eat a bite.

"Why aren't they telling us anything?" Amber whined.

Axel seemed to be holding up better. "I'm sure it'll all work out. Don't worry. Maybe we'll get elected together, you never know!"

Amber gave her a distracted grin and rolled her little ball of wax-covered cheese back and forth. "It just feels like something is wrong. What if something happened to the ballots?"

Axel shrugged, "They'd just have another election I guess, right?"

"You think?" Amber was doubtful.

"Well, no news is good news. I'm sure it will work out, even if it's a do-over." Axel closed her empty lunchbox. "I'm still hungry. Don't worry so much. Hey, maybe we'll even get something ELSE in the paper besides how amazeballs your little friends are. Would do them some good to remember they aren't the only important kids in school."

"Hmm?" Amber was only half-listening. She'd barely touched her lunch. Offering the red wax cheese wheel to Axel, she asked, "You want this?"

"Sure, thanks," Axel unwrapped the wax and bit into the soft white cheese, talking around it. "Can you imagine being announced a loser? Anyway, even that would be refreshing after all the Renegade write-ups. It's about time we got some worthy news on the front page."

"It was just two articles, Axel," Amber reminded her. "They just happened to hit a slow news day."

Axel didn't reply so the conversation fizzled out.

The cafeteria buzzed around them like a beehive as kids tried to figure it out. Who won the election? Where was Benjamin? What was going on? But no one had any answers.

Out in the recess yard, still soggy from yesterday's rain, Wren, Ivy, and Kammie didn't even bother to set up shop. The whole school seemed like it was on hold. Tense. Waiting for something to happen. Everyone just seemed to be waiting for lunch to be over, hoping an announcement of some kind would come in the afternoon.

But no announcement came.

29
THE ASSEMBLY

*T*he next morning, a schoolwide assembly was announced for 10:00 a.m. to address the election situation. But Axel Andrews looked pretty relaxed about it as she sat doodling in her science notebook, waiting for Mrs. Yang's substitute to bring them to the auditorium. Kammie watched her from the back of the classroom and wondered what she was thinking about. She'd heard Axel tell some other kids that she thought they'd just do another election if something happened to the first ballots, but Kammie wasn't so sure.

Axel had withdrawn a bit since the election. It made sense considering how invested she was in becoming a council member. Amber had just made some posters and thought up a slogan. But Axel really seemed to have plans. She'd even chatted with Kammie one day about her ideas to bring cultural events to the school, like musicians and poets from around the city. And she had never chatted with Kammie before. In retrospect, maybe Axel really would do positive things if she were on the council. She certainly had the ideas and the drive to get them done. Kammie felt kind of bad for her. She hoped she and Amber were supporting each other.

At 9:45, the kids headed down. The entire school crowded into the bleachers. Kammie couldn't find any of her friends, but she was supposed to sit with her own class anyway. She ended up right behind Axel on the bleachers, looking at the perky blond ponytail.

At precisely 10:00 a.m., Ms. Sophie entered. Quiet washed over the crowd as the principal stepped up to the microphone. She tapped it to make sure it worked, and pulled out a prepared statement.

"Good morning students," Ms. Sophie began in a clear, strong voice. "As you are aware, there has been an unexpected delay in our student council elections."

She paused for effect. Several of the kids nodded, leaned forward, or both.

"Our election proceedings have been the same for many years," she continued, "but this year is different. It is my unfortunate duty to confirm what you have probably already heard. The ballots and results have been stolen. In the history of our school, nothing like this has ever happened, and I am very disappointed in you."

Loud protesting whispers swam up from the crowd.

Kids sounded angry about the accusation. Kammie continued to quietly stare at the principal. Axel sat stone still, waiting.

"In addition to the thief or thieves themselves, surely someone knows who the culprit is. Whether this theft was a prank or something worse, someone knows something. Yet no one has come forward. As a result, we likewise cannot move forward with the elections this term. After much thought and discussion, I have decided NO ONE will be appointed to the open seats on the council this term."

The kids roared unhappily. Several students gasped. Axel suddenly shot upright in her seat, shaking her head no. She turned to say something to Tiffanie next to her. Kammie heard the word "unfair" over the din.

Ms. Sophie lifted her hands for silence and after a few seconds the students grudgingly settled down.

"I know it's upsetting, but without the official results or ballots, there is no way to certify the election. Rules are rules."

"Just ask Benjamin!" a voice called out. Kammie thought it might have been Bobby. Several kids called out in agreement.

"Proceeding based on the word of those students who did the counting would be against the election rules," Ms. Sophie replied, "it places an unfair burden on them. With no proof, they could face accusations about their honesty that they don't deserve. I'm sorry students, but I don't see another solution. This situation is unprecedented, upsetting, and disappointing. Without the official ballots or tally sheet, there is simply no other path forward. Next year, if we continue the student government program, new protocols will be established. But honestly this incident makes me question the school's ability to handle the responsibility of a student government at all.

"By the rules of the elections, a week from Monday will be the last day any changes can be made. That gives you tomorrow and all of next week to come forward. If any of you know anything about the election robbery, confess before then. If no one comes forward, my decision will be certified. At that time, I will decide if we will continue the student government program next year."

The principal waited calmly while more chatter rose up, then died down.

"I hope this unpleasantness will not detract from the dedicated work your fellow students have put into preparing for the competitions happening this upcoming week!" Suddenly upbeat, the principal wrapped up the assembly. "On Wednesday, our own Lovelace Math Machines team will face off against the McKinley Badgers, and the Saturday after that we will host a city-wide swim meet here at our pool. Our second graders are working hard to bring together an art show the same Saturday we host the swim meet, so there will be something for students of all ages. I strongly encourage all of you to attend these exciting events and show your school spirit! Let's support our school. Let's show the city what Ada Lovelace Charter School can do! Go Machines!"

The room erupted into chaos, but not the usual cheering. No more student council. Seriously? After all the good it had done for the school! As Kammie rose to leave, she saw Axel, pale and shaking, reach out to put her arm around the shoulders of another distraught girl. She was comforting her quietly. The other girl, Mikayla, wiped away a tear and nodded her thanks to Axel. Kammie didn't even remember if Mikayla had been a candidate. She paused to look for her friends. Unable to find them, she rushed off to return to class and passed Axel again. This time standing by a wall encouraging Bobby sympathetically. All of a sudden, Kammie felt like maybe she hadn't really seen Axel before. At least not this side of her. She wondered what else she had misjudged about the perky blond girl.

30

ACCUSATION

A tall boy walked into Mr. Vincent's English class that afternoon with a note. Kids looked up curiously from their reading discussion groups.

Mr. Vincent frowned when he read it and glanced at Wren and Ivy's group. Everyone followed his gaze. Wren, Ivy, Milo, and Lily looked at each other.

Mr. Vincent came over, crumpling the note as he walked. "Please go see Ms. Sophie after class," he said quietly.

"All of us?" Milo squeaked.

"Just Wren and Ivy. Carry on, you still have fifteen minutes left of discussion time." He turned to check in on one of the other groups, but most of the kids were too busy staring at Wren's group to discuss anything literary.

"What did YOU do, Ivy?" asked Lily. "I mean, Wren

goes down there all the time, but have you EVER been called to the principal's office?"

Ivy shrugged nervously. The fact that she was included in the summons made things very mysterious. She couldn't remember ever being in the principal's office. And given Ms. Sophie's reputation, she'd probably remember.

"Maybe it's a good thing," suggested Milo. "Maybe you got an award or something."

"Or your Dad got hit by a car and you need to go to the hospital." offered Lily helpfully.

"BOTH our Dads? Seriously?" asked Wren, a bit offended no one was surprised at her summons too. She hadn't been sent to see Ms. Sophie in forever, but apparently no one besides herself had noticed.

Ivy shrugged nonchalantly. "I guess we'll have to wait and see."

"I hate waiting to see," Wren scowled. "That never goes well."

The reading group didn't discuss their book much after that. No one in the class seemed to. Mr. Vincent tried unsuccessfully to refocus them on their work more than once, but the allure of a cryptic note from the principal was too juicy a mystery to ignore, especially since they were still reeling from the election news. Everyone suddenly seemed to have a ton of theories about both things.

Wren knew the rumors would start spreading after class. It had happened to her before. She sunk a little lower in her chair and scowled.

Ivy, on the other hand, now seemed convinced Ms. Sophie might give them some sort of award. After all, they hadn't done anything wrong. Why else would she be called in?

Their confusion only deepened after class as they approached Ms. Sophie's office and found Amber and Kammie already waiting outside her door. Amber sat on one end of the row of chairs, and Kammie sat at the other avoiding eye contact.

The four girls fidgeted on the hard plastic chairs as they waited. No one said anything, and Wren refused to look at Amber, even when Ms. Sophie called them in.

There were no guest chairs in the office. That was when Wren knew they were in serious trouble.

The four girls stood shoulder to shoulder in a row in front of Ms. Sophie's desk. She contemplated them over steepled fingers.

"Hey, I like the new curtains, Ms. S!" Wren said conversationally.

"Thank you, Wren," the principal replied with a cold voice. She paused expectantly, then asked, "Is there anything else you want to tell me? Any of you?"

They glanced at each other. Ivy cleared her throat, but

couldn't find her voice. After an uncomfortably long silence, Wren spoke first.

"I don't think so. I've been thinking really hard, Ms. S., honestly I have. And I just don't know what I did this time!"

Ms. Sophie's expression softened slightly. "I believe you all know Benjamin Spencer?"

Wren snorted. "Who doesn't?"

Ivy elbowed her.

"We recently had business dealings with Benjamin," Ivy added as professionally as she could. "Has he complained about the quality of our product or something?"

Kammie mumbled something about how they should have provided tech support, but no one heard her except Wren.

"You could say that," Ms. Sophie leaned forward. As you know, on election night Benjamin stored the ballots and election results in the safe you sold him. The next morning, the results were gone."

All four of them nodded, waiting for Ms. Sophie to go on. To explain why they were here in her office.

"Yeah," Wren prompted after more silence with Ms. Sophie watching them expectantly. What did she want them to say? "That's what you said at the assembly."

"I don't understand," Ivy said. "Is there some way we can help?"

"I'm afraid it's more serious than that, Miss Park," Ms. Sophie said. "I need some answers from you. I'm investigating a robbery."

"It wasn't me!" Wren blurted, trying to lighten the mood. It came out more seriously than she'd intended.

"Don't be so dramatic," chided Ivy. "No one is accusing you." She looked at Ms. Sophie. "Right?"

The principal leveled a serious look at all four girls. "Actually, Ivy, that is exactly what I'm doing. Right now, you four are my main suspects."

Shocked silence filled the room. Not even Wren said anything. Finally, in a small voice she said, "It really wasn't me."

Kammie silently scooted two steps behind Wren.

"There's been a misunderstanding," Ivy frowned. "We don't have any special tricks to get in the safe! It was specifically designed so Benjamin would be the only one who knew the combination. None of us know it any more than anyone else at the school!"

"You'll have to tell me more about that," Ms. Sophie pondered. "But even so, you'd still be my prime suspects."

Amber's brow furrowed in confusion. "But... Why would you think WE had anything to do with it?"

"You're running for a place on the council, aren't you, Miss Rosenberg?" she asked. "That gives you a motive, an interest in the results."

Amber nodded silently.

"Lots of people were running, Ms. S.," Wren couldn't help defending Amber, even though she was mad at her.

"That's certainly true, Miss Sterling. However..." Ms. Sophie reached into her drawer and placed two items on top of her desk. "These were found around the corner near the scene, further incriminating your club. Pretty compelling evidence, unfortunately."

There on her desk sat a unique, bespoke utility belt labeled "Ivy," and a purple telescoping cardstock periscope, with "SAMPLE - Property of the Renegade Girls Tinkering Club" written boldly and clearly on the side.

31

REUNION

"I thought we were doomed!" Wren leaned against the wall outside Ms. Sophie's office when they were finally released.

Thanks to Wren's long standing track record of telling the truth, the whole truth, and nothing but the truth to the principal, Ms. Sophie was willing to listen when Wren said they hadn't stolen anything.

But the motive and physical evidence were hard to deny. And it sounded suspicious to keep saying that both the periscope and Ivy's belt happened to get lost at the same time. Even to themselves. Even though that's exactly what happened.

To make matters worse, the girls realized their inventions had been used in a way they never intended. To hurt people.

Ms. Sophie had let the girls go even though there were still a lot of unanswered questions. No one knew what would happen next. But for now, the girls were free.

Out in the hall, Ivy, Wren, and Kammie stood in a small huddle, trying to think of something to say to each other.

Ivy squished her eyes closed and rubbed them with her fingers. "We'll just have to figure this out. I don't know what's going on, but we'll figure it out and fix it okay?"

Amber fidgeted with the bottom of her cardigan, standing a few steps away from the group, watching the others, refusing to leave. Her lips trembled slightly. She stared at her blue sequin-toed shoes. Wren and Ivy had their backs to her, but Kammie could see her. As she watched Amber fidget, tears began cascading down Kammie's cheeks. How had everything gone so wrong?

"Oh, Kammie!" cried Wren, wrapping Kammie in her arms protectively. "Everything's going to be okay. I promise!"

Wren leaned back, held Kammie firmly by both shoulders and tried to look her straight in her downcast eyes. "We will figure this all out and make everything okay. Okay? If anyone knows how to get out of trouble, it's me, and I will NOT let anything bad happen to you. We're a team and we've got this. Together. Okay?"

Kammie nodded silently.

Amber tried to clear her throat but it came out as a squeak.

Ivy and Wren jumped, noticing her for the first time. Wren glared at Amber, pulling a crumpled but unused tissue from her pocket and handed it to Kammie.

"We don't know what happened, but I can assure you we had nothing to do with any theft," Ivy took on a practical, business tone talking to Amber. "You didn't either, I assume? You weren't lying in there? You haven't been trying to get us in trouble have you? Because that would be pretty low, Amber."

Wren glared, a protective hand on Kammie's arm.

"Of course not!" Amber cried, her own eyes glistening with tears. "I'd never do anything like that, come on! Look, I'm sorry. I miss you guys so much! I wish I'd never gotten so mad. I wish I'd been more helpful making the safe. I'm sorry I was mean. Especially to you, Wren. I was wrong. But you guys hurt me too."

Wren narrowed her eyes. Kammie gently put her hand on top of Wren's. Amber looked at their hands, together, protecting each other, communicating wordlessly. She squeezed her eyes shut and hung her head.

"You made such cool things without me, like you don't even need me. You didn't help me with my stuff, you know? It's like you don't even care what I want, all that matters is the spy gear business," Taking a few steps towards them, Amber looked up. Wiping her eyes with the corner of her cardigan, she reached out tentatively. "But really, I guess I was jealous."

Wren's expression softened. "That's stupid."

Ivy frowned, "The safe was really hard, we could have used your help. But, come on, Amber. Did you expect us to just stop everything when you walked out? We make do with what we have. Renegade style. And we keep our promises. It doesn't mean we don't need you."

"I know," Amber nodded. "I'm sorry."

"And I'm sorry we weren't more helpful with your council stuff. I liked your slogan," Kammie blew her nose. "We've been pretty focused on making money."

"Yeah, and it was actually pretty nice to be the center of positive attention for a change, at least for me," Wren added. "I guess that was kind of selfish. Sorry about that. We could have been a lot more supportive. But, honestly, Amber? You hurt my feelings. A lot."

"I guess it doesn't matter now," Amber sniffled. "Looks like no one is going to get a seat on the council. This mess is my fault. Ms. Sophie suspects us because I'm the one with the reason to steal the election results."

"Yeah," nodded Wren. "That's true. That sucks."

Amber gasped and Ivy frowned at Wren, who threw up her hands and shrugged. "What? Come on, it DOES suck, that's just being realistic. I mean, you didn't do anything wrong, but running for council put us in a bad situation. Not to mention it was a complete surprise."

"I didn't mean to," Amber whimpered.

"The important thing now is what to do next," Ivy turned to Wren and Kammie. "I mean, however we got here, here we are, right? So what I see is that we need to figure out what actually happened; who really took the ballots and results, and how our stuff got to the crime scene. That's the only real way to clear our names."

"I could... you know... help?" Amber offered, biting her lip nervously. "I can help you figure it out. I mean, I'm in trouble too. I miss you guys. I miss the club. I want to be part of it again. Please."

The other three girls looked at each other. Ivy raised her eyebrows at Wren. Wren frowned and started to shake her head, but then closed her eyes. After a second, she nodded.

"Alright." Ivy said. "We miss you too. Despite everything, you're one of us. Once a Renegade, always a Renegade. What's the first rule of the Tinkering Club?"

"Lock the gate when you leave?" answered Wren, confused.

"No. The first rule of the Tinkering Club is that mistakes are okay. Getting mad is okay." Ivy choked up a little. "Life is frustrating, Amber. And you don't have to be perfect. We don't have to be perfect. None of us are."

Amber broke into tears and gathered them into a giant hug. But Wren stayed stiff, not fully committed. She'd give it a try, but honestly wasn't sure she could ever really trust Amber again.

Amber felt Wren's hesitation but couldn't help laughing anyway. She was happier than she'd been in weeks, despite the cloud of suspicion hovering over them all.

32

RUMORS

*W*ord had gotten out.

Everywhere Wren looked, kids seemed to be staring at her, whispering about her, suspecting her. In the morning as she walked down the subdued hallway, no one said hello.

"Excuse me," she had to prompt the group of kids in front of her locker. They grudgingly moved aside, lowered their voices, and kept talking quietly to each other.

Grabbing her science book, with its completely unfinished homework shoved inside, Wren closed her locker and trudged to class. At least Amber was back at her usual seat. Wren flopped next to her and tossed her notebook on the table, noticing Milo enter the room. He started towards Wren, then saw that Amber was back in the seat he'd been using. Turning to sit with Bobby instead, he glanced back at Wren with a worried look. She returned a helpless smile and a shrug. No one else sat near Amber and Wren.

"Well this sucks," Wren pouted.

"What?" Amber was setting out her worksheet and pencils.

"Everyone thinks we stole the results. We are pariahs. Outcasts! No one will ever speak to us again!"

Amber looked up, surprised. She glanced around. "What? What do you mean? I haven't noticed anything. Everyone's upset about the election stuff. Maybe you're picking up on that? Did you get to number nine in the word problems?"

But Wren couldn't even think about homework.

"It's not just that," Wren persisted. "Don't you see it? Everybody is talking about us."

Amber scrunched up her mouth. "I don't know, Wren. Maybe?"

By lunch recess, Amber had to admit something was up. Not even half their usual number of customers came by. And the ones that did weren't very chatty.

"Well, enjoy it," Ivy told a girl who had bought the first periscope of the day. The girl left without replying.

"See?" declared Wren. "Everyone's mad at us."

"You think so?" Kammie replied quietly, a worried look on her face. "Oh gosh, I hope not."

"Hmmm," pondered Ivy. "The whole mood of the

school has been down since the election. I doubt it's personal. Things will pick up."

But they didn't.

On Saturday, they weren't able to meet because Ivy had two games back to back and Wren's mom told her she couldn't do anything until her math homework was finished. Amber had been looking forward to meeting again, but even she had to admit they didn't need to make any more spy gear. Their inventory remained full.

By Tuesday, despite having a bulging inventory, the business was dead in the water. They wondered why they even bothered setting up.

Sitting on top of their usual picnic table in the chilly fall sun, Kammie said, "I guess that's the end of our spy business. I know I was against it at first, but it ended up being so fun. I hope we at least made enough money to pay for the microscope eyepieces."

She pulled a little spiral notebook full of numbers out of the money box.

"What's that?" asked Wren, looking over her shoulder.

"Our log book. I wrote down everything we spent and recorded every single sale."

"Why?" Wren asked.

"So we could see how much money we made of course, after factoring in all the expenses and stuff," Kammie replied matter-of-factly. "It's weird though. I wrote down every item we ever made in our inventory and what happened to it in this little book..."

"Geeze! That IS weird!" agreed Wren. "And kind of obsessive."

"No, no! Keeping records is just good business. I mean about the stuff Ms. Sophie found. That sample periscope and Ivy's utility belt are the only things we ever lost this whole time. Every item and all the money we spent and earned is all accounted for except them. Isn't it weird to be so careful with everything but happen to lose exactly the two things that ended up at the crime scene?"

"Yeah, actually," replied Wren thoughtfully. "That IS super suspicious. I bet someone absconded with them!"

Ivy had been half-listening to their conversation. "I'm still not convinced we have to close shop," she said. "We have a lot of inventory left. What are we supposed to do with it all?"

"Right, exactly," Kammie frowned at the log book. "We just spent money on new supplies to make more stuff before the election. And you know, after factoring in what supplies cost, we really didn't make that much on each sale. We should have priced stuff higher."

Amber, absentmindedly braiding a chunk of her hair, shook her head. "If we had charged more, kids probably wouldn't have bought so many things."

Kammie shrugged. "Well, we certainly didn't charge Benjamin enough for that safe. Those metal sheets and everything really added up. It was expensive to build. We may have even lost money on that."

Amber sat up straight and dropped her braid. "No way," she said stoutly. "He did so much for us! We should have given it to him for free!"

Wren frowned at her, but Kammie just shook her head sadly, "Well, I think we just barely have enough right now to get the new eyepieces. I think. They're pretty expensive."

"But stuff was selling so well!" cried Wren. "We should have enough for three sets of eyepieces!"

Kammie just shrugged.

"Money is stupid," pouted Wren.

The girls sat in disappointed silence while the recess yard churned with kids playing all around them. Kids who were once again not interested in their table.

"You know what?" Amber said quietly. "I'm worried about Benjamin."

Wren rolled her eyes. "Seriously? Benjamin is just fine. It's not like anyone thinks he stole from himself!"

"No," Amber continued quietly. "I mean. Do you think he still likes us?"

At that, Wren was silent. Ivy and Kammie looked at each other. Would Benjamin really think they'd stolen the election results? Would Gail?

Nobody asked the question out loud.

33

HEADS TOGETHER

*T*he Renegades held an emergency meeting at afternoon break.

"We have to talk to Gail and Benjamin," declared Amber. She had considered asking if she could invite Axel, but knew Wren wouldn't like it. Besides, since the assembly Axel hadn't really wanted to spend any time with Amber. Or anyone else really. Amber made a mental note to check in with her later and make sure she was okay.

"Wait!" cautioned Kammie. "We have to think it through first. What would we even say?"

Wren seemed lost in thought. She pushed her hair behind her ears and said, "You know, I don't think we lost those two items at all."

"What?" said Kammie, feeling sidetracked. "We were talking about Benjamin."

"I've been thinking a lot about how strange it is that we lost those two particular items only," Wren continued, undaunted. "I don't think we did."

"How is that relevant to what we're talking about?" Kammie threw up her hands.

"I think we're dealing with something bigger here."

"BIGGER?" Amber squeaked. "What could possibly be bigger than the whole school thrown into chaos and worse, Benjamin not liking us anymore? And Gail, too."

"I think we're dealing with a schoolwide thief. Possibly an organized ring of counteragents. Spies who are using our clever equipment to commit a series of crimes and frame us for them. It's dastardly. Throwing the school into chaos might even be their objective..."

Ivy looked at her doubtfully. Amber started to say something but suddenly Wren gasped. She jumped off the picnic table.

"What if, bear with me here, but what if it's kids from McKinley? They'll be the other team at the math meet, right? What if they're trying to destabilize our student government!?" Wren started pacing back and forth while pondering the possibilities. "They might have an agent here at school. Oh! Maybe it's Bobby! He's always up to no good. What if they think they can start beating us at school competitions if we're distracted and fighting each other about the election results?"

Ivy contemplated Wren, who put her hands on her hips in triumph.

"It all makes perfect sense! All the pieces fit." Wren looked expectantly at her friends.

Ivy forced herself to consider the possibility and was shocked to realize it wasn't as crazy as it sounded at first. The others just stared at Wren.

"That's a creative theory for sure," Ivy admitted. "I don't know about plotting some giant takeover of the student government, but it wouldn't be the first time another school pranked us. But how would we prove it?"

"Isn't it obvious?" Wren said with frustrated patience. "We have to become spies ourselves! It'll be fun!"

Silence.

Then Amber actually nodded. "You know, that actually might work. But I still don't know how it helps us with Benjamin and Gail."

"Sometimes in life, Amber, people stop liking you," Wren shook her head sadly. "That's just the way it goes. But I'm talking about an adventure!"

"You're talking about trouble," Kammie said. "Even more trouble than we're already in."

"Yes! Exactly. Trouble, adventure!" Wren's eyes sparkled. Maybe her mom was right. Maybe she wasn't a troublemaker. Maybe she was an adventurer. Maybe trouble was just another word for adventure. It was all in how you looked at it sometimes.

Kammie, however, whimpered.

"Alright then," Wren declared triumphantly, "let's solve the case and find the thief! Let's be spies!"

34

THE SPY LOLLIPOP

That night, in Wren and Trixie's shared bedroom, at the desk under Wren's loft bed, a spy wasn't getting her math homework done. She'd done a few problems, but was now stuck on number twelve. If Wren was going to be a successful spy, she'd need to go on missions, which meant leaving her house. And if her math homework wasn't complete, she'd be trapped in this bedroom forever. She had to finish this page.

Three times now, Agent Wren had attacked the equation. She knew what to do. She always knew what to do. The problem was going through the steps and filling in the numbers. Numbers that would escape from her brain only to be replaced with images of nefarious counteragents from McKinley. The evil spies in her head were busy kidnapping President Benjamin and disrupting the stability of

Ada Lovelace Charter School. Only the brave girls from the Renegade Spy Agency (International) could save him!

She wasn't getting anywhere with problem number twelve.

Her fourth try floundered when a sudden knock at the door broke her flow of concentration. She growled.

"What!?" she barked at the door, expecting yet another "check in" from her mom. "Still working!"

To her surprise, Trixie's little head poked in as the door cracked open, her eager little face framed by messy pigtails.

"Hi," said Trixie timidly, stepping into their room with a small pink bag poorly hidden behind her back. She knew she was supposed to stay out when Wren was doing her homework, but had grown impatient waiting outside. With a quick glance to make sure Mom hadn't seen her, she closed the door.

"I made something spy," the younger girl told Wren in a loud whisper. "For you."

"Really?" Trixie was adorable when she was an alternative to homework. "What is it?"

Trixie proudly offered her the small pink bag. Several straws stuck out of the top at odd angles.

Wren pulled out a straw. A large blob of what looked like gum was stuck to one end. On the other end was a large handmade pom-pom of fluffy yarn. Wren couldn't immediately see how the pom-pom was stuck to the straw, but it stayed in place as she waved the straw around.

She looked at her sister quizzically.

"It's the spy lollipop!" A giant smile burst forth on Trixie's face. "You guys didn't make them so I did. Mom helped me make pom-poms. And maybe helped figure some of it out, but mostly it was all my idea."

It did indeed look like a lollipop. A lollipop made from spare parts. Wren couldn't figure out what made it a spy gadget, though.

"How does it work?"

"Well," Trixie whispered loudly and leaned closer. "When you want to make sure no one sneaks into your room, you use it to see if someone sneaked into your room."

She took another lollipop and pulled the pom-pom off the end. A doubled-up pipe cleaner about four inches long slid out of the straw, wrapped around the pom-pom. The pipe cleaner ends twisted around themselves, stiff and fat. It wouldn't come out of the straw on accident. Wren took the contraption and poked the pipe cleaner back into the straw. It went in easily. Another large glob of white gum was stuck to the other side of the pom-pom.

Wren patted Trixie's cheek. "Spy lollipop, cute. Thanks. I've got to finish my homework, though."

Trixie's face crumpled as Wren set the mess of spare parts aside. "No, really, Wrenny. It works."

Trixie wasn't going to let her focus on problem twelve

until she explained her invention. Plus, honestly, Wren was kind of curious. Glancing at her homework with resignation, Wren turned to her sister. "Show me?"

Under the side drawer of the desk was a cubby with a door. Trixie pulled it open. A messy stack of papers fell out and Trixie kicked them underneath the desk.

First, she stuck the straw to the door. Then, she stuck the pom-pom side of the lollipop to the inside floor of the cubby using the gum, with the pipe cleaner sticking up towards the straw. Next, she closed the door, threading the pipe cleaner into the straw carefully as the door closed.

Finally, she opened the door, pulling the straw off the pipe cleaner, which remained stuck to the floor of the cubby.

Immediately, Wren saw it. If you knew to look while you opened the door, you could see if the pipe cleaner was still in the straw. If it was, no one had opened the door. If the pipe cleaner was not in the straw, someone had opened the door before you. It was elegant and simple.

"But," Wren pondered, "won't the gum dry out?"

"Gum? Ew! That's not gum. It's that sticky stuff Dad puts under his toys to keep them from falling down in an earthquake."

"Earthquake putty?" Wren asked, picturing her Dad's collection of action figures lining the shallow shelf above his desk.

"Yeah, okay. I found it in Mom's drawer. It's sticky and

good for sticking stuff." Trixie quickly added, actually whispering this time, "Don't tell Mom."

Wren opened the door on the desk cubby, watching the spy lollipop come apart. Then she closed and opened it, seeing the device had been tripped. She reset it and closed the door again, repeating a few times in wonder.

"Wow, Trix... you made this? For me? Why?"

Trixie looked up at her sister.

"Because you were sad."

Wren reached out and gently smoothed a rogue bit of hair behind Trixie's ear.

Abruptly, Trixie smiled and slipped quickly and quietly out of the room like a secret agent.

Wren returned the spy lollipop to the pink bag with a soft smile and carefully placed the whole bag in her backpack.

Then she turned back to number twelve.

SPY LOLLIPOP

MATERIALS:

Regular sized straw
Pipe cleaner
Earthquake/adhesive putty
(Tape works if you don't have putty)
Large Pom-Pom

Optional: Make a fake door out of cardboard again

Cut a piece of straw about 3 inches long. Stick it to the inside of a door, down near the floor

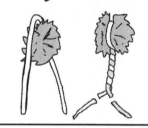

Wrap the pipe cleaner tightly around the Pom-Pom, and twist it down a few inches. Trim ends

Poke the pipe cleaner into the door straw, and stick the Pom-Pom to the floor

DOOR CLOSED   DOOR OPENED   ENTRY DETECTED!

When the door opens, the pipe cleaner pulls out of the straw. Open it slowly and check the straw to see if someone opened the door! Now brainstorm ways to make it better. How would you improve it? Or think up a whole new invention!

35

HYPOTHESIS

"*D*id you hear Ms. Sophie at the assembly last week?" Wren asked the group at lunch. "McKinley will be here for that math competition TONIGHT! It's our big chance to catch them in the act. Do some spying!"

"What, McKinley again?" Kammie facepalmed.

"Benjamin is on the math team," Amber pointed out.

"That's true," said Ivy. "We could actually talk to him outside of school. Otherwise, between school, the math competition, and the big swim meet on Saturday, I don't think we'll get a chance to talk to him until next week. And by then Ms. Sophie might have closed down the whole student government."

"Then we have to go!" Amber stated. "It's settled."

Kammie sighed but didn't say anything.

The girls passed around Ivy's cell phone to call their

parents and let them know they'd be staying late to attend the competition. Not surprisingly, none of them objected to their daughters' sudden interest in high-level mathematics. Especially since Wren had finally finished problem number twelve. Amber's mom even offered to take them all out to Amber's favorite taqueria afterward and drop the kids off at their houses if it meant encouraging Amber's interest in math.

The competition would be held in the library a few hours after the end of school. The Renegades planned to head there right after their last classes let out and finish up some homework until the school emptied. Once the halls were clear, they'd break out some leftover spy gear and do a little *Sarah and Simon* style reconnaissance, which was the spy way to say "spying."

Wren was so excited that she finished all her homework in record time, sitting in the quiet library with the others as the librarian set up chairs. With an hour left before the competition, it was time to investigate.

The halls of Lovelace were eerie, silent. The dim lights and quiet footsteps of the girls tiptoeing down the hall soothed Wren. She felt alert and more clear headed than she was used to feeling in school. The candidate's posters had been removed and recycled, replaced by a few educational ones in sedate colors. Wren stopped in front of a big poster that said SCIENTIST vertically down its length with the virtues of a scientist extolled across each letter.

Like "per**S**istent", "**C**urious", "**I**nquisitive" and so on. Standard, boring, inspirational school fare. Wren ran her hand over it. So much less obnoxious than the glitter and balloon covered monstrosities that used to pack the halls.

Even the sedate, creamy wall color was diluted with the lights off and no fluorescent flickering or buzzing. She wished she could go to school like this every day. But they weren't here for classes now.

An auburn head poked around the corner of a fourth-floor hallway, then pulled back.

"Amber! Use the periscope for heaven's sake!" Ivy's voice came from somewhere down the hall.

"Whoops," Amber replied. "Coast is clear."

Wren and Kammie approached from the other end of the hall, moving carefully. Kammie looked over her shoulder as the girls met up.

"Second floor is clear. Did you check the third floor?" Ivy asked.

"Does the crow fly at midnight?" Wren replied with a wink.

"What does that even mean?" Ivy asked.

"... yes." Wren replied with a sigh. "You're a terrible spy."

Amber looked at the lockers nearby.

"You know," she whispered, half to herself, "that's Benjamin's locker right there. That means they found our sample periscope somewhere around here."

Kammie glanced around and pointed to a corner very close to the locker. "There, maybe?"

Ivy stayed by Benjamin's locker while Kammie, Wren, and Amber moved around the corner into the other hallway. Wren nudged Amber, pointing to the periscope in her fancy, black velvet spy sash utility belt.

Amber hid behind the wall next to a poster that said ENGINEER, telescoping her periscope to full length. Looking around the corner with it, she could clearly see into the locker as Ivy opened the door. She was hidden, but close enough to see the contents of his locker in detail. After a few seconds, Ivy moved and blocked Amber's view of the two fiction books stacked on Benjamin's shelf. Amber tried to see around her, curious what he was reading.

Suddenly Ivy was coming towards her. She pretended Ivy was Benjamin and she didn't want him to see her. Where would she go? Two steps down the hallway was a water fountain alcove. She zipped into it, flattening against the wall under a big TECHNOLOGY poster.

"Where did she go?" she heard Ivy ask quietly.

Her friends appeared around the edge of the alcove.

"Well, I guess we know for sure how the thief got the combination without being caught," Wren mused. "They could watch Benjamin unlock it and he wouldn't suspect a thing. Dastardly!"

"If I were a bad guy, I totally might have dropped the

periscope by accident." Amber mused. "You came around the corner pretty quick. I had to think fast to hide."

"Or the counteragent might have dropped it on purpose," Wren disagreed. "To frame us."

"Either way, this alcove is the only place to go in time. It's a little dark though," Amber pointed out, pulling a UV necklace from under her shirt.

The purple light it gave off when she squeezed the sides wasn't much, but it was all she had. She waved it around as she tried to step out of the alcove. But Wren held up a hand, trapping her inside.

"Wait a minute," Wren was looking at the wall. "Shine that light back on the wall. Right about here."

Amber obediently pointed the flashlight where Wren pointed. Three large, fluorescent numbers and letters appeared like magic.

"Hey," she said, "What's that?"

"I don't know, but it's written in our ink," mused Wren. "A2X. What's A2X?"

"That's the combination to my safe. What are you doing in my locker?" said Benjamin Spencer, arms crossed, as he and Gail Mendez stood next to his open locker. They looked from the girls to the open door and back again accusingly. "Besides breaking and entering?"

"This isn't what it looks like!" squealed Amber.

36

CAUGHT IN THE ACT

"Oh?" Gail scowled. "And what, precisely, is it then?"

Amber, wilting beneath Benjamin's disapproving stare, said nothing. It was Wren who finally spoke.

"We're trying to catch the counteragents. We're here to protect the school. We're on your side!"

Kammie groaned quietly.

"Counteragents? What are you talking about?" Gail's face remained stony. "You know, I really took a chance featuring you Renegades in the paper. Front page, even! Not only have you messed things up for the whole school, you've hurt my reputation as a reputable journalist."

Gail turned her back and closed Benjamin's locker.

"We didn't steal anything," Ivy floundered. "In fact, we got stolen from too! We think something big is going on."

Gail continued to look at them with scorn, but Benjamin's expression changed slightly.

"What do you mean?" he said.

"Don't listen to them, Ben," Gail said. "We caught them red-handed. They're trying to weasel their way out of trouble again."

"No," Benjamin replied. "I want to hear what they have to say,"

"Sorry," Amber finally spoke. "We were going to close your locker. We were just trying to figure out how someone might have gotten the combination to the safe."

Benjamin nodded slowly, "And?"

"From this corner here, using a periscope, we could see straight into your locker. Then Ivy tried to catch me doing it, and I was able to hide in this alcove. Then we found the invisible writing in the alcove, probably written so the spy had your combination for later without using a scrap of paper that might be found or lost. Or maybe they just didn't have anything else handy, I don't know," Amber shrugged. "You guys came by before we had a chance to..."

"Cover up your trespassing?" asked Gail.

"No! Well, yes. Maybe," Ivy stammered. "But we proved our theory. The utility belt and the periscope that were found had been stolen from us."

Gail just nodded sarcastically. "Sure. That doesn't sound fishy at all."

Wren, digging in her backpack, pulled out the pink bag of spy lollipops.

"Hey," she said, "we have a present for you guys. Whether you believe us or not, the agents from McKinley will be here tonight for the math thing and Saturday for the swim meet. I don't know who, if anyone, is working with them here at the school, but I want you to be on your guard."

Gail rolled her eyes, but the beginnings of an amused smile played at the corners of Benjamin's mouth.

"And what's this?" he asked, taking the gadget Wren offered him.

"It's a spy lollipop. My sister made it. Long story, but I think it can actually help us."

Benjamin spun it around, fluffing the yarn pom-pom, the smile trying to force its way through. "How old is your sister?"

"Six, almost." Wren waved her hand dismissively, "But here's the point. You can use it to see if someone has been in your locker."

The other Renegades and Gail were watching their discussion. Wren hadn't had a chance to tell anyone about the spy lollipops yet and they were all intrigued.

"May I?" Wren motioned towards the now-closed locker. At Benjamin's nod, she opened the door and showed everyone how the lollipops worked.

Benjamin was impressed and more than a little amused. "Runs in the family, eh?"

Wren didn't seem to notice the comment.

Gail looked at her phone. "Look, this is all cute, but we have to get to the library. The math competition is about to start, and I couldn't find anyone else to report on it. And we are not leaving you ladies here alone to get in anyone else's locker."

Ivy nodded, "We're actually here to go to the competition."

"Oh!" Benjamin perked up. "You like competitive math? You're in for a treat!"

"Math is great," Amber mumbled, blushing. "Go team!"

As they hurried towards the library, Wren clarified, "We're here to protect the school. And its government!"

Benjamin chuckled as Gail rolled her eyes again.

"My personal bodyguards," he joked, giving Amber a quick poke in the ribs with his elbow.

She giggled.

"I can't believe you," Gail groaned at him.

There weren't many chairs, but then there weren't many people at the competition either. It was just a small, informal practice match.

Gail pointed to a row and said, "All of you will sit here, where I can see you."

She sat herself in the last seat of the row while Benjamin went to join his team.

The Renegades had never been to a math competition before and didn't know what to expect. Wren had a tough time focusing on it, and soon pulled *The Morocco Mystery* out of her backpack and buried her nose in the thick book.

Kammie started out mouthing the math problems to herself, trying to figure them out with the competitors. She was getting into it when one of the McKinley Badgers suddenly left the room. She pointed the suspect out to the others. They were trying to figure out how to get past Gail to tail him when he came back, drying his hands on the type of paper towel found in school bathrooms. He wouldn't have had time to make it to the upper floors and still go to the bathroom to wash his hands. A dead end.

Ivy was busy sizing up the McKinley Badger kids. Even though the Badger from the bathroom hadn't had time to steal anything, she took his picture with her phone anyway. She could always pretend she was on the newspaper staff if anyone asked. Gail scribbled a few notes here and there, but mostly played games on her phone. Amber, though, surprised herself. The competition was actually really fascinating. She had no idea how interesting math could be. Benjamin, with the other Lovelace Machines on the team, seemed awash in pure joy.

Eventually the event wrapped up. Gail interviewed Benjamin as they headed towards his locker to grab his jacket and books. The Renegades started towards the door, where Amber's mom planned to pick them up.

The evening had been a complete waste of time. Anti-climactic. They had hoped for a lot more adventure, but all they'd accomplished was to get in more trouble with Benjamin and Gail. They were horrible spies. Wren had secretly wished for a high-speed chase or at least a chase on foot through the school hallways, where she'd break out her karate skills and take down the counteragent at the last minute. Was that too much to ask?

Suddenly, Amber started pawing through her backpack.

"Darn it!" she cried. "I can't find my periscope! Uuuugh! That's all we need. Another periscope lost by Benjamin's locker!"

She dashed off.

The other girls rushed after her in hot pursuit. She went straight to the water fountain on the fourth floor. Her periscope was where she'd dropped it when Benjamin had appeared earlier, forgotten on the floor of the alcove.

She picked it up as the others arrived, but Wren was distracted by sounds coming from around the corner. From the direction of Benjamin's locker.

Benjamin and Gail were there, looking shocked and worried. They both glanced up as the Renegades cautiously approached. Gail's attitude changed dramatically.

"I guess I was wrong," she turned back to something inside the locker.

he safe last night. Maybe I should bring it Saturday
swim meet. I wonder if they'd try to break in again."

here's only one thing we can do," decided Wren, half
self. "Set a trap to catch them!"

o her surprise, the others agreed. Except for Kammie,
vas busy biting her nails.

Dh! Gail! Can you write a story about Benjamin
ing the safe back?" Amber suggested. "That might lure
in."

Gail held up the copy of the week's paper she'd
ght to the table. "They just came out today. Even if I
d get something online tonight, I doubt the informa-
would get around fast enough to be any use. How else
d we..."

'I got this," said Wren, and gave a giant theatrical gasp
group of kids stumbled by, pretending not to stare.
u're using the SAFE AGAIN? That's GREAT,
NJAMIN!"

"What are you doing?" asked Gail suspiciously.

More kids passed by unusually close, slowing down
I pretending not to eavesdrop. Wren winked at Gail.
'hat's that Gail? BENJAMIN is bringing the SAFE to
CHOOL on SATURDAY? Wow!"

Kammie looked around frantically as the roving kids
oved on, talking amongst themselves. "Shhhh, Wren, my
rs! Why are you talking so loudly?"

"No, Kammie. The SAFE we made for BENJAMIN!

"What's going on?" Wren asked excitedly, moving closer. "What happened?"

"Well, thanks to this spy lollipop of your sister's, we've discovered you all were telling the truth," Benjamin pointed to the mess of straw and pipe cleaner stuck to the door inside his locker. "Looks like I was broken into again. And we know for sure it wasn't you, because you were with us the whole time. I think you're right. I think something big is going on."

37

ALL IS NOT LOST

"Someone infiltrated the school, and we will find out who it was," Wren declared for the third time.

It was lunch recess on Thursday, the day after they'd proven themselves innocent. To Gail and Benjamin anyway. To the surprise of pretty much every single student at Lovelace Charter School, Benjamin Spencer and Gail Mendez sat in the recess yard chatting with the Renegade Girls Tinkering Club. Groups of kids passing by stumbled over each other as they caught sight of the unexpected grouping.

Axel, who had spent the last few days eating lunch alone at the corner picnic table, had looked up as Benjamin approached the recess yard. When she saw him and Gail sit with Wren and her friends after dropping off the week's newspapers, Axel burst into tears and ran off. Amber was

worried about her. She hadn't been a[l] with Axel outside of school for almos[t] since the assembly. Amber shook her he[ad] be even more upset than she was. She [ ] try again after school.

"Why can't we just go tell Ms. S[ ] handle it?" Kammie pleaded.

"I don't know what she could do right [ ] you guys off the suspects list," Benjamin [ ] "And she only has our word for even that [ ] anything. We need some proof."

Gail had her notebook out, pencil read[y] what we know."

"What we know is that someone is [ ] McKinley to steal our school secrets," Wren [ ]

Everyone ignored her.

"We know someone wants to break in[to] locker," Ivy offered. "They stole spy gear fr[om] and maybe even to frame us. We DON'T k[now] they might have stolen around the school."

"Agreed," Gail wrote it down. "We know [ ] election information, but we don't know why."

"And they struck again last night," Wren [ ] This time Gail wrote it down.

"They also know the combination to the [ ] Amber.

"Good point," Benjamin agreed. "Althoug[h]

"What's going on?" Wren asked excitedly, moving closer. "What happened?"

"Well, thanks to this spy lollipop of your sister's, we've discovered you all were telling the truth," Benjamin pointed to the mess of straw and pipe cleaner stuck to the door inside his locker. "Looks like I was broken into again. And we know for sure it wasn't you, because you were with us the whole time. I think you're right. I think something big is going on."

ALL IS NOT LOST

"*S*omeone infiltrated the school, and we will find out who it was," Wren declared for the third time.

It was lunch recess on Thursday, the day after they'd proven themselves innocent. To Gail and Benjamin anyway. To the surprise of pretty much every single student at Lovelace Charter School, Benjamin Spencer and Gail Mendez sat in the recess yard chatting with the Renegade Girls Tinkering Club. Groups of kids passing by stumbled over each other as they caught sight of the unexpected grouping.

Axel, who had spent the last few days eating lunch alone at the corner picnic table, had looked up as Benjamin approached the recess yard. When she saw him and Gail sit with Wren and her friends after dropping off the week's newspapers, Axel burst into tears and ran off. Amber was worried about her. She hadn't been able to get in contact with Axel outside of school for almost a week now. Ever since the assembly. Amber shook her head. Poor Axel must be even more upset than she was. She would just have to try again after school.

"Why can't we just go tell Ms. Sophie and let her handle it?" Kammie pleaded.

"I don't know what she could do right now, except take you guys off the suspects list," Benjamin explained gently. "And she only has our word for even that. We don't know anything. We need some proof."

Gail had her notebook out, pencil ready "Let's go over what we know."

"What we know is that someone is working with McKinley to steal our school secrets," Wren replied.

Everyone ignored her.

"We know someone wants to break into Benjamin's locker," Ivy offered. "They stole spy gear from us to do it, and maybe even to frame us. We DON'T know what else they might have stolen around the school."

"Agreed," Gail wrote it down. "We know they took the election information, but we don't know why."

"And they struck again last night," Wren tried again. This time Gail wrote it down.

"They also know the combination to the safe," said Amber.

"Good point," Benjamin agreed. "Although I didn't

have the safe last night. Maybe I should bring it Saturday for the swim meet. I wonder if they'd try to break in again."

"There's only one thing we can do," decided Wren, half to herself. "Set a trap to catch them!"

To her surprise, the others agreed. Except for Kammie, who was busy biting her nails.

"Oh! Gail! Can you write a story about Benjamin bringing the safe back?" Amber suggested. "That might lure them in."

Gail held up the copy of the week's paper she'd brought to the table. "They just came out today. Even if I could get something online tonight, I doubt the information would get around fast enough to be any use. How else could we..."

"I got this," said Wren, and gave a giant theatrical gasp as a group of kids stumbled by, pretending not to stare. "You're using the SAFE AGAIN? That's GREAT, BENJAMIN!"

"What are you doing?" asked Gail suspiciously.

More kids passed by unusually close, slowing down and pretending not to eavesdrop. Wren winked at Gail. "What's that Gail? BENJAMIN is bringing the SAFE to SCHOOL on SATURDAY? Wow!"

Kammie looked around frantically as the roving kids moved on, talking amongst themselves. "Shhhh, Wren, my ears! Why are you talking so loudly?"

"No, Kammie. The SAFE we made for BENJAMIN!

He's going to bring it BACK for his SATURDAY SWIM THING," Wren practically shouted to the next group of kids.

She sat down next to her friend and threw an arm over her shoulders.

"Kammie," she said, "if anyone knows how fast rumors spread around here, it's me. When I lose my temper in math class, the whole school knows by lunchtime."

Gail was looking at Wren with new appreciation. "I may have underestimated you."

"Everybody does," Wren nodded.

"That gives us one day to think up a way to catch the thief," Ivy's face got businesslike. "And I have an idea brewing. I just have to convince my mom to give me another lift to the electronics store!"

38

SPY TRAINING

*A*lone in the Greenhouse on Friday night, Wren was thinking about how to catch a thief. They had until Saturday evening to prepare their trap. The problem was, they really sucked as spies. Their last spying attempt had proven it.

What they really needed was training. Spy training. And Wren was the only one who seemed to understand its importance.

"I've got this," Wren told herself confidently.

Ivy and Kammie were working together on some electronic invention doodad, which was integral to their plan. Benjamin was splashing around somewhere practicing for his swim meet, and Gail was... Wren wasn't really sure what Gail was doing but it sounded important. Completely unrelated, but important. So Gail would be no help. And she still didn't trust Amber.

It was up to Wren now. She'd have to take charge. She'd have to be their Agent Zero to prepare them for the mission. And they needed to be a lot more prepared than last time they infiltrated the halls at school. They had to level up their spy game.

Wren finally understood there was no point in squishing down her sense of adventure. After years of failing at being someone else, Wren was ready to be herself. And it felt good. The spy gadget business might be over but, if nothing else, it had shown her that all she needed to do to keep her brain pointed towards the adventure side of trouble was to give it something creative to do. Then the power, the drive, of her brain was unstoppable. And that drive was exactly what the Renegade spies needed now.

In front of her, the table was covered with sketches, bowls, and a bottle of clear glue. She pushed a leftover bottle of UV ink, some yarn, and a few different rolls of masking tape to the side as she poured some pre-mixed borax water into the glue waiting in a bowl in front of her. Plunging her hands in, she squeezed and squished, but the slime was way too sticky. Frustrated, she dumped some raw borax powder into the bowl and tried to mix it in. Everything clumped together.

She dumped the whole messy glob into the trash. Useless. She needed someone better at science to help.

Someone who actually paid attention in class and remembered the little details.

After washing her hands with soap, she flopped back onto the stool and looked over her pages of hastily scribbled spy training plans. She ran over everything in her head. It was a good idea, but she couldn't do the training at her own house. It was too small and messy. And Trixie would get in the way.

She needed someone with a bigger house to make it all work. Someone who could help her with the science. Someone with access to a printer.

She knew someone exactly like that. She just had to choose to trust them. And trust was hard for Wren. All day every day, people told her she was trouble, she was bad. If she trusted those people, she'd have to believe what they said. So Wren had learned to be selective about the people she trusted. And wasn't very good at giving second chances. But, maybe some people were worth the risk. Maybe.

Wren took a deep breath and let it out slowly. Leaving everything in a pile on the Greenhouse table, she slowly made her way into the kitchen. Reaching hesitantly for the phone, Wren dialed a number she hadn't dialed in weeks. A number she knew by heart, but wasn't sure she'd ever dial again. The phone rang a few times. Then, just as she was about to hang up, someone answered.

"Hello?"

"Hi, Amber. It's Wren. I need your help."

39

RENEGADE SPIES, INTERNATIONAL

*A*mber's grandmother lived at Amber's house, in an apartment downstairs. Ivy had been there a few times, so when a coded message full of clues had been slipped under her door, she figured it out quickly.

As Ivy walked up to the apartment, she was surprised to see Kammie already standing outside. They'd finished the invention last night, and she hadn't expected to see Kammie again until the swim meet that evening.

Kammie looked relieved when Ivy arrived, "Did you get a secret message too? I hoped I figured it out right. I just got here, but was starting to worry I was wrong. I'm glad you're here. What should we do now?"

"I guess we knock," Ivy shrugged.

Ivy reached over and gave the door a confident rap. Kammie cringed.

The door swung open immediately. Amber, dressed in

black yoga pants and a long sleeved black thermal ski shirt, stood in front of them. Her hair was pulled into a tight bun. She stepped aside and motioned them into the apartment. It was dark except for some dim light coming from a glob Amber held in her hand. As Amber closed the door, the glob became the only source of light, and it was a bad one.

"What's going on?" Ivy asked as Kammie looked around with wide eyes.

Amber led them into the hall closet without a word. She closed the door behind the girls and suddenly the light flipped on, blinding them.

"Hey! Ouch!" yelped Kammie, throwing her hand over her eyes. She blinked her vision clear, then looked around.

Everyone just barely fit in the empty closet. Kammie, Ivy, and Amber turned to Wren, who stood in the corner with her hand on the lightswitch.

"Sorry!" Wren apologized. "A little brighter than I'd planned! These glasses are really dark."

An adult-sized gray fedora hat, too big for her head, threatened to swallow Wren's whole face. Only a pair of enormous sunglasses kept it up. Her chin was barely visible above the dusty black leather trench coat Wren had dug out of the back of her mom's closet. She looked like a secret agent straight from a cheesy movie.

"Wren," Ivy crossed her arms impatiently. "Why are we in a closet?"

"Because we are SPIES, Junior Agent Ivy. We need privacy. Secrecy. We need to remain vigilant, always," Wren clapped her hands together. "Alright, junior agents! Welcome to the Renegade Spy Agency International!"

"International?" asked Kammie, "Really?"

"We have branches in several major metropolitan areas around the globe," Wren confided with a smile.

"You are so weird," giggled Kammie.

Wren tipped her fedora in acknowledgement and continued. "Welcome to spy training! If we're going to catch the counteragent and save the school, we must prepare. To protect our student government we need to do more than collect information and make wild guesses. We need to catch this thief! WHO'S WITH ME?"

The others stared back at her silently. Wren frowned beneath the sunglasses. She motioned insistently with her hand. Kammie cocked her head to the side. Wren motioned again. Finally the others relented with some half-hearted applause.

"Good enough," declared Wren. "Agent Sciencemaster and I, Agent Feline, have prepared a spy training obstacle course. Your mission, should you choose to accept it, is to find the radioactive material without setting off any alarms, and disarm the bomb."

Amber held up the glob in her hand. Now that the light was on, Ivy and Kammie could see it was a glob of clear slime. It wasn't glowing anymore.

"What's that?" Ivy asked.

"That is the radioactive material!" Wren said, "Go with it."

"Actually," clarified Amber, "it's glow-in-the-dark slime. Remember the slime we made in science class? This is the same stuff, but we used clear glue and mixed in a bunch of glow powder before adding the borax water. Isn't it amazing?"

Amber held up the glow slime with one hand and let it sag into the other with a slow, non-Newtonian style drip. "There's more in Grandma's apartment. You have to go through the laser maze and disarm the bomb before you can get it!"

Kammie suddenly looked nervous. "Laser maze?"

"Don't worry, it's not real," Wren smiled. "But it'll be challenging in the dark. Amber and I have done it a couple times!"

Wren slipped off her hat, sunglasses, and trenchcoat dramatically. Underneath she was wearing the same kind of tight, flexible black outfit that Amber wore. Except her shirt looked a size or two too small in the sleeves. She handed two UV flashlight necklaces to each of them. "Since we have so many extras anyway, you can use two so you have enough light. But, umm, be careful okay? Amber's grandma took out most of the breakable stuff but we promised her we wouldn't, you know, mess anything up. Ready?"

Bewildered, Ivy and Kammie took their necklaces, and Amber and Wren slipped some on from their pockets. Wren flipped off the light again while Amber opened the door.

The girls stepped out of the closet gingerly as their eyes adjusted to the dark. The living room was down a long, thin hallway. Ivy started walking towards it but Wren stopped her. She pointed her UV flashlights down the hall.

A criss-crossing series of glowing lines appeared all the way down, like lasers criss-crossing a tunnel in a spy movie.

"Wait," Kammie whispered. "Are those... lasers?"

"RIGHT?" Wren burst out. "They look JUST LIKE lasers, huh? Actually, we dipped yarn into the UV ink and taped it to the walls. The flashlights are floressanting them. Isn't that BRILLIANT?"

"Fluoresce," Amber smirked. "The flashlights are fluorescing them. The yarn is now fluorescent so it's fluorescing, it's reflecting the UV light from our flashlights. I know they're a little hard to see because our flashlights aren't very strong, but it's a really cool idea. The point is that you have to crawl through the maze without pulling any of the yarn from the walls. It's fun! Try it."

Ivy went first. Wren watched, making a loud, unpleasant MEEEP whenever Ivy touched some yarn. After the third MEEEP, Ivy stood up, hands on her hips, turning angrily to Wren. As she stood, she ripped some yarn lasers completely off the wall.

"FAIL!" Wren called. "Reset!"

Ivy went back to the beginning as Amber slid gracefully through the maze to tape the yarn back in place. Amber continued to the other side.

"See guys?" she called. "You can do it! Wren and I both got through it successfully."

"I'm glad you guys are getting along again," Ivy confided to Wren as Kammie carefully wove her way through the lasers.

"Yeah, me too," Wren admitted. "Never underestimate the trust-building properties of lasers and radioactive material. Your turn! Don't mess up this time."

Ivy smirked at her and turned to the maze again. She sized it up and moved forward with determination.

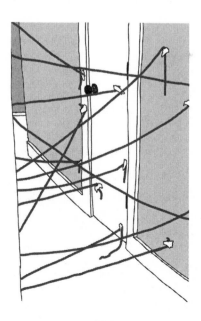

It didn't take long before all the spies-in-training stood on the other side of the laser maze, shining their flashlights around the dark room. Ivy had to admit it was pretty fun once she'd made it through.

"Where's the bomb?" Kammie called, shining her light under the couch.

"Wait, it's not there?" Wren squeaked. "It's supposed to be under the couch."

"Granny will kill me if it pops and gets radioactive goo all over everything!" Amber looked around frantically. "It must have rolled away!"

Wren paused. "You know it's not really radioactive, right?"

"Uh, yeah. I helped make it," Amber scowled at her. "But that'd probably, honestly be better than a bunch of slime on her rug."

The girls searched around but couldn't find it.

"Wait! I know!" Wren declared. "Teamwork!" Gathering the Renegades together, Wren had them point all their flashlights together in one place. The extra UV light finally picked up a faint glow under the dining table.

"I looked there!" Kammie said. "I guess my two lights weren't strong enough."

Wren crawled under the table for the bomb: an inflated balloon they'd filled with glow-in-the-dark slime. "You have to defuse this bomb to get the objective."

"Got a pin?" Ivy asked.

"No!" Kammie objected. "That would explode it and spread the goo everywhere! We have to deflate it without popping it. I'm on it. I do this all the time with balloons. I hate it when they pop."

Kammie held the stem of the balloon tightly, well above the knot. With her other hand, she grabbed the knot and held the secure section out to Ivy, who carefully cut between the places Kammie held tightly. The knot came off but since Kammie held the stem, the air didn't escape. Kammie gently, slowly let the air out of the balloon from the open stem. When the balloon was completely deflated, she took the scissors and cut away the floppy latex, exposing the glowing slime.

"You underestimate yourself, Kammie," Wren nodded with approval. "You'd make a fine field agent."

Kammie blushed. "I'd still rather be a cryptologist."

"Well," Wren began handing out little business card-sized pieces of cardstock. The cards said *Renegade Spy Agency, Clearance Level* 3. "Whatever kind of spy you want to be, I'm glad you're on my team."

40

EVENING EVENTS

*T*he gym roared as athletes from several schools stroked and crawled and splashed their best efforts for their teams. It was Saturday evening, and the swim meet was packed. Parents and classmates cheered and waved homemade signs and scarves in whatever colors represented the school they supported.

The pool, rare for a big city school, was in the gymnasium across the street from the main Lovelace building.

In contrast to the gym's bright lights and activity, the main school building sat dim and still. The second grade art show was just closing down, and the building would be locked up soon. Ms. Sophie gazed at her school fondly, breathing deeply in the cool evening air. It had been a rough few weeks for her students between the run-up to the election and what happened afterwards. She'd been worried about her kids from the moment Benjamin

Spencer had told her about the election theft. But tonight, she was proud to see that many of them showed up for both events. It was important to her to attend in person as well.

But honestly, she just wasn't fond of sporting events. They were too loud and too bright. Sometimes the plucky little sixth-grader, Wren, reminded her of herself as a girl. She could also feel overwhelmed by large crowds and lots of activity. She preferred less stimulating activities. The art show had been calmer. Now, however, a lot of those families crowded into the gym to cheer on the Lovelace Machines' swim team, creating even more chaos. Ms. Sophie enjoyed the school spirit, but needed a break.

So here she was, sitting on a bench outside the gym, breathing the quiet San Francisco fall air. Gazing across the street at her beloved school as it sat sleeping. Dark and quiet.

Then she saw it. A light in a window!

And it was gone.

Ms. Sophie blinked and looked again, waiting. And there it was again, in another window.

It looked like someone was in the school with a flashlight, moving around on the fourth floor!

Ms. Sophie stood up and smoothed her jacket. Who was wandering around her school at this hour? The art show was on the first floor, on the other side of the building and the few families left had no reason to be up in the middle school area. Especially not on this side. It was unlikely to be the teachers, either, since they'd be busy closing down the show. So who could be wandering around up there? Whatever was going on, she'd find out. Someone was going to be in serious trouble when she caught them. Her kids had been through enough.

Scowling, Ms. Sophie punched the crosswalk button with a stiff finger, even though there were no cars this time of night. Rules were rules.

As soon as the walk sign lit up, she strode across the street rolling up her sleeves, on her way to deal out some justice to whomever was messing around on the fourth floor of her school.

41

READY? ACTION!

A flashlight's round beam of light moved along the row of lockers on the fourth floor, pausing briefly at each number. The beam stopped on one near the corner. Quiet feet tiptoed towards it.

Around the corner, all four Renegades pressed themselves against the wall near the drinking fountain alcove. Ivy had her periscope in her hand, but it was too dark to use it. Amber held a small digital camera with her hand capped over the light that indicated the flash was ready.

They'd managed to sneak in unseen, thanks to their dark clothes. And probably the spy training had helped warm them up too. They'd moved through the silent halls as one unit, Ivy on point (which was the spy way to say in the front) and Kammie bringing up the rear. Wren and Amber had kept watch as Ivy and Kammie installed their invention in Benjamin's locker. With his permission, of course. Then they'd scurried to relative safety near the alcove to wait.

Just in time, too, because they'd heard a door close somewhere down the hall. Then they'd seen a flashlight's beam wavering around.

Now in place and following their plan, all they could do was wait and hope nothing unexpected happened.

They waited silently as the flashlight's beam continued to get closer. Suddenly, it paused on Benjamin's locker number and the quiet footsteps quickened. Ivy raised a periscope to watch the approaching light.

Everything was falling into place exactly as they had planned. Soon, they'd have answers. And proof. They were just waiting for the sign before Amber could get photographic evidence of the thief's identity. Just another minute...

Then, in one split second, everything crumbled like a poorly-baked cookie. Ms. Sophie herself emerged from the stairway straight across from them. Luckily, she was wearing practical shoes that didn't make a sound.

But that was where their luck ran out. They were trapped against the inspirational engineering poster— caught red-handed. This time it would be impossible to explain what they were doing there. The trap hadn't been sprung yet! If Ms. Sophie scared away the real thief, they'd never clear their names. They were so close!

Ivy's mind raced frantically trying to figure out what to

say to the principal. In the periscope, she could see the shadow holding the flashlight was almost at Benjamin's locker. Just a few more seconds!

Ms. Sophie, angry, disappointed, and serious all at once, reached for the light switch.

The Renegades panicked. Three of them froze in terror. But not Wren.

Wren, thinking fast, gestured silently, frantically. She caught the principal's attention and put her finger to her lips, pointing and gesticulating wildly.

Ms. Sophie stopped, fascinated. She watched Wren's bewildering, silent movements with confusion. She tilted her head, trying to figure out what the girl was trying to say. Narrowing her eyes as Wren continued to fling her arms around silently, Ms. Sophie finally gave up and took a deep breath to say something stern and adulty.

As she opened her mouth, a loud, piercing screech sounded. It took everyone a split second to realize the ear splitting noise was coming from around the corner, not from Ms. Sophie's open mouth.

The Renegades sprinted around the corner as a shadowy figure stood rooted at Benjamin's open locker. One shadowy hand held a fat manila envelope, and the other held open the door to Benjamin's safe, which was tucked securely on the shelf inside the locker.

Amber's flash went off as she captured the thief's

image, illuminating a blond ponytail just as Ms. Sophie flipped on the hall lights.

The electronic wail drowned out the buzz of the lights as they flickered to life. Another sound joined in the racket, coming from the figure in front of the open locker door. It was a voice they all knew.

Axel Andrews, throwing the manila envelope into the safe and covering her ears as the screeching from the locker continued to buzz, screamed, "WHAT ON EARTH IS THAT NOISE?!?!"

42

THE TRAP IS SPRUNG

*A*xel leaned against the wall, shoulders slumped. Her bangs covered her face as she stared at her feet. A small tear dropped from behind the hair onto clenched hands.

As soon as Ms. Sophie had turned on the lights, Ivy had rushed over and turned off the buzzer, much to everyone's relief. Her invention, a tripwire alarm, was set to go off as soon as the safe opened. It had done its job. She grabbed the manila envelope from inside the safe and passed it to Ms. Sophie.

Ms. Sophie checked inside.

"It's all there," Axel said softly. "All the ballots, the tally sheet, everything. Afsheen and Tyrone are our new council members. I lost."

Amber gently put her hand on Axel's trembling shoulder.

"I... I'm so sorry, I didn't know it was you," Amber said quietly. "Or we would have done things differently."

Axel wiped the back of her hand across her eyes before lifting her head.

"I get it," she replied, flicking her hair out of her face and straightening up. "I'm not mad."

Wren snorted.

"Explain yourself, Miss Andrews," commanded Ms. Sophie in her cold, powerful voice.

Axel wilted again. "I'm sorry! I didn't mean to. Honest! It was an accident."

"Accident?" Wren yelled. "ACCIDENT? You ACCI-DENTALLY almost destroyed the whole student government? You ACCIDENTALLY stole stuff? You accidentally committed a CRIME? Are you working for McKinley?"

"What?! No! I... what?" Axel looked completely confused. "I'm not working for anybody! I only wanted to look. I was so sure I was going to win, I just deserved a peek before anyone else saw it. It wasn't a big deal, or anything!"

"Except the part about breaking and entering and robbery and safe cracking and spying. Oh! And framing us!" Wren spat angrily. "No big deal! You ruined our business!"

Ms. Sophie put a hand on Wren's shoulder. "That's enough. I will deal with this situation."

Wren huffed and crossed her arms, leveling Axel with a withering stare. But she stopped talking.

"Look, Ms. Sophie," Axel tried again. "When I saw I didn't win... I mean, I didn't want anyone else to see. Nobody likes me!"

"Have you tried not being a royal jerkface?" Wren blurted.

Ms. Sophie gave Wren a warning glance. "You apparently came in third." Ms. Sophie told Axel sternly. "I hardly think that qualifies as no one liking you. And it certainly doesn't justify your actions."

"I know." Axel sniffled. "I didn't realize how much it would affect everybody. It just sort of happened. I've been trying to put everything back, but Benjamin took the safe home . And I thought if I just stuck them in the locker loose, somebody might take them." Realizing the irony of what she just said, Axel flashed Wren a look that dared her to have another outburst.

Wren kept her mouth closed for once.

"You will stay right there, young lady. I'm not finished with you," Ms. Sophie pointed a finger at Axel, rooting her to the wall, and turned to Ivy.

Ivy still stood at the open safe, removing the tripwire alarm. As Ms. Sophie approached, she offered her the invention and began explaining it.

Meanwhile, Amber was trying to get up the courage to

ask the most important question on her mind in the most casual way possible, pretending she didn't care.

"So..." she asked Axel, "just wondering... how many votes did I get?"

"You didn't win," Axel said diplomatically.

"Well, I guess it's not true NOBODY likes you, huh?" Amber tried to laugh it off. "I mean, you did better than me apparently."

"That's true," Axel sniffled. "I did beat you."

Amber's smile faltered.

"Sorry!" Axel quickly amended. "I mean... Good job! Yay..." she shrugged with a helpless smile. "You actually came in eighth. Which, considering everything, was at least not horrible. I'm just sorry you got wrapped up in the whole thing when your friends almost got in trouble. That was not what I planned when I took those toys from your little club."

43

BETRAYAL

"You did it on purpose?" Amber looked at the girl desperately. "But, Axel... why? Why didn't you tell me? Why didn't you DO something? You knew I was in trouble, and you were going to let me get blamed. For something YOU DID. How could you?"

"Hey, it wasn't on purpose!" Axel shrugged. "Look, I needed that spy stuff so I could see the results. They weren't really using them. They were just samples." Axel waved a hand dismissively towards the other Renegades. "I was going to give them back, but I dropped them when I hid from Benjamin. Then it all got weird. I'm sorry, Amber. I really am. I did what I could. But you know how it is when you have a reputation to keep up. I have such good ideas for the council. I was so close! With a redo I'd get in for sure and then it would all be worth it. For the good of the school! I couldn't risk that. I've been trying to put the stuff back and keep the government open. That would have cleared you."

"No, it wouldn't. People would still think I stole it. You were going to sacrifice my reputation so you could get on the council. If not this year, then next year. Weren't you?"

"It's not like you were going to win anyway," Axel glowered at her. "I would have thought you'd understand. I just couldn't risk exposing myself. You know how it is."

Amber backed away, shaking her head. "No, no I really don't. My life is about more than just one election."

"Well it all worked out okay, didn't it? Even if you guys hadn't gone all spy, I still would have put it back."

"But... people would still think we did it. And... you're okay with that. Axel, that's ... that's just wrong."

Axel shrugged. "I guess you don't understand after all. You don't understand what it's like. What do you want from me?"

"I wanted to be your friend. But that's not how friends act."

Ivy brought the tripwire alarm over to them. Axel cringed away from the mess of string and electronic parts.

"What is that thing anyway?" Axel looked at it like it was a sack full of spiders.

It was a wooden clothespin taped to a big C cell

battery. The SUPER LOUD electronic buzzer Wren had found at the electronics store was also taped to the side of the battery with two wires dangling from it. One wire was attached to the negative end of the battery with electrical tape, and the other end was wrapped in aluminum foil and secured to one of the jaws of the clothespin with copper tape. A length of yellow wire was attached to the inside of the other jaw with copper tape too, so when the clothespin was closed, the two wires and all their metal wrapping would press against each other. The other end of the yellow wire went from the clothespin jaw to the positive side of the C battery, taped down with even more electrical tape. They'd used a lot of that stretchy plastic tape!

The two metal-wrapped jaws of the clothespin clamped around a bit of paper keeping the metal jaws from touching each other. A string dangled from a hole in the paper.

"See!" Ivy explained proudly. "This alarm is wired in a circuit, just like our UV flashlights. The paper keeps the circuit from being completed, so the buzzer doesn't buzz when it's in there. If the paper gets pulled out, the power can go from the battery, along the wires, through the metal of the copper tape, to the buzzer, and back to the battery, completing the simple circuit. And of course, we know what that means..."

Ivy looked at their faces, but the blank looks from Axel

and Amber told her that actually they didn't know what it meant when electricity completed a circuit with a buzzer attached. Luckily Wren joined them in time to hear the explanation.

"Why, Ivy, that means the buzzer will go SCREEEECH!" She yanked the paper from the jaws, allowing the two jaws to connect. Sure enough a piercing wail immediately started.

Ivy glared at her and squeezed open the jaws. As soon as the metal jaws weren't touching each other, the screech stopped. She stuck the paper back in and let the jaws clamp back around it.

"I hooked the clothespin to the safe, and taped this string to the door. When you opened the safe's door, Axel, it pulled the paper out and closed the circuit. Our trap was sprung and the buzzer went off. We were going to take your picture and a video to bring to Ms. Sophie, but she showed up in person. I don't even know what she's doing here, but I guess it turned out okay. Well, for us anyway!"

Axel stared at the alarm with its wires and buzzers and batteries and copper tape. She looked dizzy.

"BAM!" Wren beamed. "You've been simple circuit scienced!"

"Nerd," Axel replied.

Just then, they heard footsteps smacking up the staircase. Benjamin burst into the hallway, his hair was sopping wet and he was struggling to get his shirt on - obviously straight from the pool.

"I saw the lights!" he panted. "I came as fast as I could. What's going on?"

Amber began to giggle uncontrollably.

TRIPWIRE

MATERIALS:

Electrical tape
Some string
A wooden clothespin
A 3V *pre-wired* electronic buzzer (make SURE it's 3 volt!)
Copper tape with conductive adhesive (and/or aluminum foil)
A spare, insulated wire, 3-4 inches long, ends stripped
A C or D sized battery (can use coin cell in a pinch)
Two non-insulated metal paperclips
A pair of wire strippers
A piece of cardstock, cardboard, or heavy paper

Tape a paperclip to each end of the battery with electrical tape. Make sure it's tight!

TIP: WRAPPING EACH PAPERCLIP IN FOIL HELPS IT CONDUCT ELECTRICITY

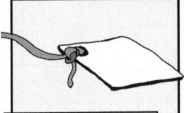

Cut a square of heavy paper and tie a string about 3 feet long to it.

The buzzer buzzes when electricity flows in a circle from the battery to the buzzer to the battery again. BZZZZZZZ! The paper stops the flow and stops the buzzer from buzzing.

Without the paper, the jaws clamp together and the electricity flows.

ALARM

Now the tricky part; attaching the wires correctly. To make sure all connections are snug, wrap them with electrical tape.

Wrap one end of the spare wire around the paperclip on the positive (PLUS) end of the battery. Wrap with electrical tape.

Adding foil or copper tape before the electrical tape is optional, but helps for a good connection.

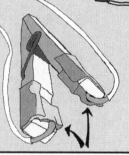

Secure the black (NEGATIVE) wire from the buzzer to the paperclip on the negative (MINUS) end of the battery. Wrap tightly in electrical tape.

Use copper tape to secure the remaining stripped wire ends to the inside of each jaw of the clothespin.

Attach the alarm on one side on the inside of a doorway or in a hallway, near the floor. Clamp the paper in the jaws and pull its string taut across the walkway, securing it to the other side. When someone walks in, they'll trip the string, pull out the paper, and the buzzer will BUZZ!

THE ENGINEER

*a*s Ms. Sophie marched Axel away down the staircase, everyone else could feel the weight of accusation and suspicion falling away.

Wren excused herself to get a drink from the water fountain while everyone else chatted happily.

She paused to look at all of them from a distance before heading back. Benjamin gave Amber a quick hug. Amber blushed furiously while Ivy and Kammie smirked at her. Even Kammie was talking to Benjamin, so Wren assumed he must officially be their friend now.

Their friend. Her friends. Her team. She smiled at the group.

On the wall by the corner, the ENGINEER poster caught her eye. She read the words: crEate, iNnovate, desiGn, buIld, learN, rEason, sharE, pRoblem solve! Thinking back over the last few weeks, Wren remembered all the innovative inventions she helped design and build with her creativity. She had researched and learned about locks and UV light and spies to find answers. She'd reasoned through the logical steps of the engineering design process, and reasoned her way through other hurdles. She worked with her team, leading when she needed to and following sometimes too. She'd shared ideas and listened to ideas. But mostly, over the last few weeks, Wren had solved problems. Including some of her own, maybe.

She suddenly realized why the poster had gotten her attention. It described her, Wren Sterling. Those were the qualities not of a troublemaker, but of a good engineer. Well, a good engineer and maybe just a smidge of trouble-maker too.

cr**E**ate

i**N**novate

desi**G**n

bu**I**ld

lear**N**

r**E**ason

shar**E**

p**R**oblem solve!

"And I heard she was getting expelled!" Wren claimed, leaning back in her chair and tossing a square of cardboard towards the glass Greenhouse ceiling. Looking up, she could see clouds through the branches of the neighbor's enormous tree as they rolled swiftly past in the bright blue sky.

"No way," Amber said as Wren caught the cardboard and tossed it up again. "I don't even think she's getting suspended."

Wren caught the cardboard and turned to look at Amber. "Well that figures. Axel can talk her way out of anything. Grown-ups always believe her. I don't know what you ever saw in her that was good."

"Oh, come on," Amber said. "Nobody got hurt or anything. She just made a mistake, did something dumb. She's actually kind of nice."

Wren just rolled her eyes, "I would have been expelled."

"At least she's not on the student council," added Kammie. "And since she's not allowed to run for it ever again, I guess she never will be."

Wren started poking little holes in the cardboard with her pencil. "Hey, where is Ivy?" Wren grumped. "We've been waiting forty-five minutes."

"She's coming right after practice," Kammie replied.

"I thought soccer was over for the season," Wren frowned.

"Yes, but now it's basketball," Kammie explained patiently.

"Always something." Wren pulled out her math homework and looked at it briefly, tossing the cardboard was much more appealing. Her copy of *Sarah and Simon: The Morocco Mystery* peeked out of her bag so she pulled it out and slid it across the table to Kammie. "I'm done with this if you want to borrow it."

Amber looked over at Wren's worksheet, then glanced back at her own.

"Is that what you got for number six? I must have done something wrong." Amber rechecked her work. "Oh! I forgot to carry the two. You're right, as always. Hurry and get to number thirteen. I'm not sure about that one."

Wren smiled at Amber and bent over the worksheet with renewed purpose, tossing the cardboard square onto the floor. Her pencil had just started skritching away when they heard the creak and slam of the side gate. Ivy careened into the backyard, tossing her basketball bag next to the Greenhouse door before bursting inside.

Kammie was focusing the microscope on a slide she'd made of her own blood by poking her finger with a thumbtack.

Ivy joined Kammie at the microscope and took a peek through the eyepieces. Kammie cringed as Ivy's sweaty armpit hovered near her face.

"Whoa! Look at that," Ivy breathed. She switched on the microscope's backlight, rotated the lens, and focused again. She leaned back and made room for Kammie to look at the new detail.

"I wish we had that other set of eyepieces," Ivy sighed. "I'd like to see blood at two thousand times magnification."

Amber sighed, leaning her head on her hand. She picked up Wren's discarded *Sarah and Simon* book. "All that work," she said, flipping mindlessly through the pages. "All that trouble."

"I know, right?" Kammie pouted. "But after subtracting the cost of materials for the tripwire and what we spent on the inventory we never got a chance to sell, in addition to everything else, we barely have more in the treasury than we did when we started. Nowhere near enough for the replacement parts."

"Money is so annoying," Wren agreed, laying down her

pencil. She pushed aside her incomplete worksheet again. "I seriously cannot believe we sold all that gear, and then spent everything on stuff we didn't even intend to buy."

Ivy rubbed her nose. "We lost sight of our goal."

"We lost sight of a lot of things," Amber agreed.

Ivy reached out for a pipe cleaner and started twisting it. Without thinking, she bent the pipe cleaner into one oblong loop, and then another. As her hands worked the fluffy wire, an idea took shape. She twisted three more equal sized loops around the center, with the ends of the wire in the middle. Moving quickly, she folded a green pipe cleaner in half. She twisted the ends of the loops around the middle of the green wire, then twisted the two sides of the green wire around each other all the way down. Bending the end into a flat, circular base, Ivy placed the pipe cleaner flower on the potting table between Amber and Wren.

"Fluffy!" exclaimed Wren, gently petting the flower's looping petals.

Amber smiled, then leaned over and gave Wren an impromptu hug.

"We'll just have to think of something else to make that money," Ivy said. "We've got this. Together. Time to tinker up, ladies. Let's get started..."

46

TEMPLATES

Downloadable and printable templates and other resources available at: www.RenegadeGirls.com/projects

Cipher Code
Wheel Template

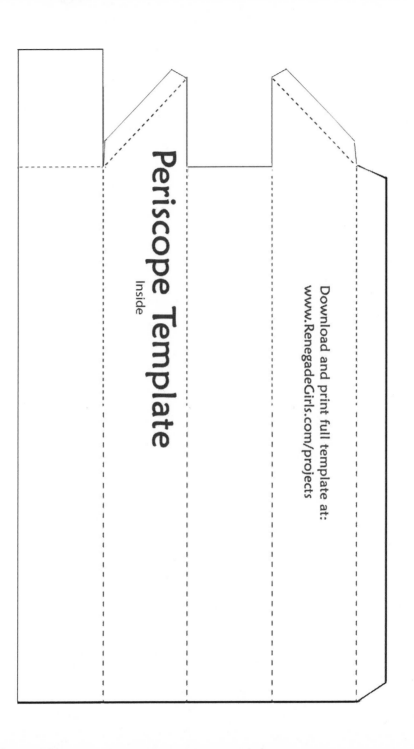

Periscope Template

Inside

Download and print full template at:
www.RenegadeGirls.com/projects

Periscope Template

Outside

GLOSSARY
KNOW WHAT YOU'RE TALKING ABOUT

A

Absconded - to run away so you don't get in trouble

Absentmindedly - without thinking about it

Amorphous - not having any real shape, like a blob

Anticlimactic - an ending that is supposed to be exciting, but is actually just disappointing

Arduino - hardware (physical) and software (virtual) electronics platform that's easy to use and relatively inexpensive. It's made for artists, hobbyists, designers, and as a way to make interactive environments

Atoms - the building blocks everything in the universe is made out of. Sort of like Nature's LEGOs. They are extremely small and made up of electrons, protons, and neutrons

B

Bespoke - made for a specific person, or made for a special order

Binocular lens - the viewing part of a binocular microscope. Some microscopes only have one eyepiece lens. Those are monocular. Binocular means there are two eyepieces, one for each eye, so it's a lot easier to see. Binocular microscopes tend to have more magnifying power than monocular ones

Blacklight - a light source that emits mostly ultraviolet light. It's not very visible, but will illuminate fluorescent objects

Breadboard - a board with a lot of tiny holes that's used to test or prototype electrical circuit

Button batteries - also called coin cell batteries. They are small, round, powerful batteries that look like buttons or quarters, commonly found in cameras, lights, car remotes, and other things. Button batteries can be dangerous if swallowed, so be careful with them!

C

Cardstock - a stiff, heavyweight kind of paper, like what business cards or greeting cards are made from. It's stronger than printer paper but not as strong as cardboard

Capacitors - a tiny electronic device that holds an electric charge, like a small, lightweight, battery that charges quickly and is built into an electronic device

Carabiner - a strong metal clip shaped like a D with

a hinged side. They can be snapped onto things and only release when the hinged side is pushed in. Rock climbers use them to snap onto their protective gear

Cipher - a series of defined steps (a procedure) used to convert plain text into or out of a secret message

Coin cell batteries - the same as button batteries above

Conductive/conductivity - able to transfer something like heat, electricity, or sound from one place to another. Metal is a good conductor, which means electricity can move easily through it. Plastic is a poor conductor (it's an insulator, which is the opposite of a conductor) so electricity doesn't move through it very well. Human conductors keep things moving, too. A train conductor is in charge of moving trains, people, and cargo around efficiently and a music conductor is in charge of keeping the music in an orchestra flowing on time and in sync.

Copper tape - a thin strip of adhesive copper with a paper backing. Copper is a great conductor (see above) so copper tape can be used in electronic projects. It's also found in gardening stores to keep slugs away from plants but for electronics projects, copper tape should have conductive adhesive (garden copper tape commonly does not).

Corrugation - a wavy, ridged surface designed to give a thin material extra strength

Counteragent - a spy on the other side

Criteria - rules for evaluating or testing something, such as a list of requirements or guidelines

Cryptologist - a scientist who studies hidden information. Usually someone who makes up or tries to figure out codes

D

Dastardly - cruel, nasty, or wicked, but usually intended in a kind of funny way

Decipher - to figure out or follow the procedure to turn a coded message back into plain, readable text

E

Entourage - a famous or important person's group of servants, assistants, or anyone who helps them. Often used to mean a group of friends surrounding one specific friend.

Etch - to draw permanently on a hard surface by cutting partway into it, specifically with acid or a sharp cutting tool

Exfoliating - a beauty process meaning to remove dry, dead skin

F

Fedora - a fancy, old fashion hat with a wide brim hat that used to be commonly worn by men. They are often seen in spy movies over a pair of sunglasses and the upturned collar of a trenchcoat.

Fluorescent - a substance that glows (gives off light) when exposed to a special kind of light or certain kinds of radiation

Force - the push or pull on an object

G

Glass blowing - the art of molding extremely hot melted glass into shapes by using several techniques, including puffing air into it

Grappling hook - a hook with one or more hook parts that's attached to a rope. Grappling hooks are thrown and hooked onto far away objects to either bring the object closer to the thrower or allow the thrower to swing, climb, or lower themselves down.

Guerilla fighter - people who fight in an armed conflict, like a war, but aren't members of a police or military force

I

Incognito - in disguise, to keep someone's identity secret

Irritant - something that causes soreness or sensitivity. An annoyance that might hurt but not seriously damage.

K

Key fob - the small device with buttons used as a key in modern cars and other locked devices. A key fob can open doors, start cars, or set off an alarm.

L

Leads (electrical) - the wires coming out of an electrical device that allow electricity to enter and exit the device.

LED - stands for Light Emitting Diode. A device that shines (emits) light.

M

Matter - a generic word scientists call all the stuff in the universe. Anything that takes up physical space.

Microbiology - a branch of science that studies microscopic organisms, or, living things so tiny you can only see them with a microscope. These tiny living things include bacteria and viruses.

Molecules - the smallest bit of a thing that is still that thing. Scientists say it is the smallest unit of a substance that retains all the properties of the substance. A molecule is made of atoms, but the atoms aren't the thing. An H_2O molecule is the smallest bit of water that is still water, but it is made up of smaller things: one oxygen atom and two hydrogen atoms.

Monstrosities - something that's gigantic or ugly, like a monster

Morosely - really sad or gloomy

N

Nefarious - naughty or wicked, evil

Non-Newtonian fluid - a fluid substance where the viscosity (thickness, the rate it flows) changes when force is applied, like when it's hit

O

Obsolete - out of date, no longer used

Opaque - can't be seen through, he opposite of transparent

P

Parallel - two things that never meet, usually lines or planes. Parallel lines are the same distance from each other all the way down, so they could be infinitely long and never touch

Pariah - an outcast, someone other people don't want to have around

Proof of concept - an exercise or test that proves a theory, concept, idea, or invention works as intended. The phrase is often used interchangeably with prototype.

Prototype - working model of an invention

R

Radioactive - something that gives off dangerous energy when its atoms break apart

Raspberry Pi - an inexpensive computer the size of a credit card. It plugs into a TV or monitor and you can program it with a standard keyboard or mouse, but it's not very powerful.

Reconnaissance - the spy (or military) word for spying. Getting more information or checking out a situation before taking action.

Reputation - what people think of someone. People can have a reputation of being nice, or mean, or silly. Sometimes a reputation is true and sometimes it's not, or it may only be true sometimes

S

Sash - a long piece of decorative fabric, like a tie. It's usually worn around the waist or over one shoulder and across the body as part of a uniform or as part of a fancy outfit

Sebum - the oil everyone's skin produces naturally. It can protect and lubricate skin and hair but too much makes skin oily

Sedate - mellow, calm

States of matter - the different ways matter can exist, the various physical properties of matter

- Solid - something with its own shape, like ice
- Liquid - something without its own shape, like water
- Gas - something without its own shape that can move freely in the air, like steam
- Non-Newtonian fluid - something between a liquid and solid that changes. It sometimes has its own shape and sometimes it doesn't, depending on the forces acting on it, like quicksand.
- Plasma - a gas with a lot of energy, like lightning

T

Telescoping - sliding one part of an object into another to make it longer or shorter

Throwies - little magnetic lights made by attaching an LED to a coin-cell battery and a magnet. Technologically minded street artists like to throw them onto metal surfaces where they glow until their battery runs out.

Taqueria - a Mexican restaurant that specializes in tacos and burritos

Toggle switch - a switch that has a small lever to push one way or the other to turn something on and off, like a lightswitch

Toxic - poisonous, dangerous

Trajectory - the imaginary path a moving object travels along

U

Utility belt - a belt used to carry important equipment in easy reach. Items are usually held to the belt with loops, pockets, or velcro.

UV light - ultraviolet light. Human eyes can't see ultraviolet light itself, but it makes fluorescent objects glow.

V

Viscous, viscosity - how fast or slow a liquid flows, how thick or thin it is. A liquid with low viscosity will flow fast, like water. A liquid with high viscosity will flow slowly, like mud. A non-Newtonian fluid will change its viscosity depending on what's happening to it, or the forces that are acting on it.

W

Welding - gluing pieces of metal together by melting parts of them or using other melted metal.

Z

Zen - being calm, like meditating

ACKNOWLEDGMENTS

Before writing this book, I rarely looked at the Acknowledgements. I knew it took a village to write a book, but what I didn't know was how honestly appreciative and deeply humbled I would be for all the invaluable help along the way.

With that in mind, I'd like to send loving thanks, in no particular order (because how do you rank infinity?) to:

My amazing husband who brings me tea, or wine, depending on the time of day, and believed in me when I didn't, and never complained (out loud) about the light from my laptop late at night.

My son, brilliant and loving, always ready with a cuddle and an ego boost whenever I'm feeling low, who never once complained about my writing a series starring his sister.

My daughter, my muse, number one fan, original test

reader and occasional editor, schooling me when "no middle schooler would ever talk like that!"

My Mom, who spent hours and days and more days editing not one but two different versions of this book. She gifted me with her brilliant critiques and insight and made my book so much better.

All three of my sisters, women in STEM careers, inspiring young girls everywhere by launching rockets to Mars, saving puppies from cancer, healing broken bones and saving lives.

My Dad, reading through a book written for 9-12 year old girls countless times and doing anything he could to help out.

My wonderful In-Laws, my belle-mère and beau-père, for their loving support, feedback, and stepping up to handle the kids and dishes so I could write.

To name a few names, thanks especially to

Vicky for co-founding Renegade Girls, and being as vibrant and brilliant as you are kind and hard-working. Juliana for writing dates, focus, oat milk chai, and friendship as well as letting me borrow inspiration from her kids (and thanks to those kids of course!). Shaheen for helping me get everything started, and for such great help along the way. Briony for inspiration and enthusiastic support. Renee for her love and support. Solène for being such an awesome inspiration. Sarah for swooping in and adding her brilliance to my confused hacks. And everyone else who has gone on this journey with me.

And I especially want to thank the thousands of Renegade Girls and Boys who've shared their creativity, fun, and amazing tinkering prowess at the Renegade Tinkering summer camps and after school programs. You are the reason we exist, you are the future. Use your voices and brains and persistence to help each other and save the world!

ABOUT THE AUTHOR

Terri Selting David's difficulties in geography manifested in 1996 when, refusing to believe that Colorado was "West", she travelled more West until she couldn't get any West-er. San Francisco provided the perfect destination to pursue her love of technology, making stuff, art, and storytelling as a digital character animator. She spent over a decade making video games, film, television, and even a comic book. But tech wasn't always the most welcoming place for a woman, no matter how talented and passionate. And her three brilliant sisters, all in STEM careers including engineering and oncology, faced similar challenges in their fields. Once her children came along, she was filled with a passion to make the world a better place for them, especially in the tech world.

So in 2015, she teamed up with her friend Vicky and founded the Renegade Girls Tinkering Club enrichment program to do what she could to make STEM a more welcoming place for the girls of the future. A few years later, she blended her background in storytelling, her digital skills, her art skills, and her experience writing curricula to write the Renegade Girls Tinkering Club novels, with a goal of bringing her projects to a wider audience and provide positive role models for girls facing the unique challenges of pursuing a love for technology, science, engineering, and math during a fragile time in everyone's life: middle school.

She lives in San Francisco with 2 rowdy children and a fabulous, brilliant husband who brings her tea every night.

ALSO BY TERRI SELTING DAVID

Check out all the adventures of the Renegade Girls at:

www.RenegadeGirls.com

And visit Terri's author website at:

www.TerriSeltingDavid.com

-

If you enjoyed The Renegade Spy Project, please leave a review!

Made in the USA
Columbia, SC
23 October 2020